Repositioning the Lent Dike – Excavations for the new secondary channel in full swing, the extended Waal bridge has just been completed, 2015

Scheveningen Boulevard – The brand new boulevard was immediately crowded during its first summer, 2013

How the Dutch Work with Water

Beyond the Dikes

Marinke Steenhuis
Paul Meurs

nai010 publishers

Depoldering Noordwaard – The old creeks in the Biesbosch become visible again as they flow during high water levels, 2016

Contents

Water World Water Land

The Netherlands is indisputably a land of water. We are a delta, the place where rivers reach the sea. The terminal point of those meandering forces. The force of water made our landscape beautiful, accessible, fertile and lucrative. Here, at the intersection of waterways, we built our water cities. It was the culture of living with water that defined how we learned to manage risks and uncertainties, turning them into strengths. Wherever I am in the world I hear people talking about the Netherlands as a land of water, where water is culture, like the blood in our veins, part of our consciousness from our youngest years. A country characterized by swimming certificates, trade, canals, landscapes and cities. Dikes and floodplains. Dunes, beaches and islands. The Netherlands is a land of water, part of it below sea level and most of it prone to flooding.

Sea levels are rising and the earth is warming up faster than ever before. Climate change acts like a magnifying glass for disasters that already exist. Storms and other extreme weather events are having a greater impact, dry periods are lasting longer, islands are disappearing, and coastal cities are sinking. Effective water management is therefore crucial: its failure is fatal, resulting in poverty, financial catastrophes, migration and wars. The situation worsens by the day; improving it following neglect becomes ever more expensive.

The global search is on for smart solutions to these overwhelming problems, creating an appreciation for the effectiveness of the Dutch approach to water management. We have been living with water for thousands of years. We built a land on the principles of flood protection and quality, and our system of water governance was in place long before the Netherlands existed as a kingdom. The Netherlands has a duty to itself and to the world to use its knowledge and expertise more than it currently does to address the worldwide water challenge. With world leaders in water technology, urban delta management and maritime technology, robust knowledge institutes, creative engineers and designers, and centuries of experience and innovation in water management, the Netherlands has a special responsibility. We have the world's most effectively protected delta. And we continue to invest below sea level. This book provides an insight into the best examples of how we have dealt with the dangers presented by the water.

To cope with the future, we must embrace it completely. This complex challenge requires a profound examination of what 'building with nature' really means. And we need to persevere in our principles of flood protection and high quality in an approach based on systems, nature and the water. Designing and developing all of the world's complex deltas requires constant innovation. Things can and must be done differently. Yesterday's solutions and approach will not meet the challenges of tomorrow. History is an inspiration but it is not a blueprint for success. We must dare to think through what the future will demand. Globally, we now have literally all the expertise we need to formulate a truly integrated and inclusive approach to deal with these very complex problems. We have the means to develop and implement these solutions. We know each other, we can work together globally, we have set future goals through the Sustainable Development Goals and the Paris climate accord (COP21). And our ambition is to innovate, to change, to make a difference. In short, we have no excuses, nor can we lose any more time.

It is actually very simple: where there is a will there is a way. And if we really want to achieve change and we harness political and social forces, we can – in view of today's global dynamic – have a real impact. We will then improve not only the lives of billions of people but also of cities and deltas, forests and rivers, creating a genuinely better world. And, seen from a different perspective: we will turn powerlessness into integral and inclusive opportunities that, based on strength and quality, will create real safety.

Henk Ovink
Special Envoy for International Water Affairs
Kingdom of the Netherlands

Employees of the Watermanagementcentrum Nederland (WMCN) in consultation, 2012

River widening Overdiepse Polder – Recreation on and next to the Oude Maasje near Waspik; the construction of new terps can be seen in the background, 2012

Below Sea Level

The Dutch live with water. Water has brought the country economic prosperity through fishing, trade and ports, and water is a source of enjoyment: the Dutch sail, skate and swim – most children have a swimming diploma. Living in a water-rich environment is exciting but it is also risky. Sixty per cent of the country is vulnerable to flooding and nearly a third of the Netherlands is below sea level. Major European rivers empty into the sea here, and two mighty forces – rivers and the sea – are decisive for the continued existence of the Dutch delta. The rising sea level is affecting the lower rivers, and their capacity is also under pressure from increasing rainfall in the Alps. Add to this the rising number of heavy downpours and growing urbanization, and the challenge ahead is obvious. In itself, adjusting to new circumstances is nothing new: for thousands of years the delta has been adapted to the water's constantly changing behaviour. The water system spreads out through the Netherlands like veins in a circulatory system, creating the network that connects its individual parts. The water system is like a family in which previous generations live on. The polders, lakes, rivers, canals, ditches, dikes, mills, locks, dams, pumping stations and barriers not only determine the identity of the landscape and cities but also shape the Dutch national identity. Cooperation and the ability to learn as well as opportunities for experimentation and innovation are qualities that are fine-tuned through interaction with the water. The Netherlands' engineering skills, drawing on the collective efforts of a wide range of disciplines, have turned the country's water expertise into an export product. It is no coincidence that it was our delta that produced the economic and cultural glory of the Golden Age, the era when the polder windmill allowed drainage of hundreds of polders. New land was created on which a prosperous farming-based urban economy thrived in the west of the country: a green metropolis, connected to the world's seas.

The proverbial struggle against the water has assumed epic proportions in history, with human inventiveness being put to the test time and again. A disaster was always the spur to heroic interventions. The Zuiderzee flood in 1916, for example, led to the construction of the Afsluitdijk and the Zuiderzee polders. Following the devastating North Sea flood in 1953, when 1,800 people drowned, the impressive Delta Works were constructed to protect the delta through a system of storm surge barriers and dams that closed off the open sea arms from dangerous storms and high water. The lessons we learned from the Delta Works have been indispensable for recent projects such as expanding the port of Rotterdam into the sea (Maasvlakte 2) and dealing with the major rivers (Room for the River). After the Second World War the south-west of the Netherlands was still a relatively isolated area in socio-economic terms due to poor land access. The construction of the dams and barriers catapulted the South Holland and Zeeland islands into the modern age. But above all else, the construction of the Delta Works created landscapes that had not previously existed: beaches *on* the Brouwersdam, splendid routes across the series of dams, new nature reserves and traffic roundabouts *on* the water. This combination of technology and landscape design has already existed for a long time. The great twentieth-century Dutch masters are not painters or poets but hydraulic engineers and landscape architects.

Benefiting from the ever-increasing water expertise, the delta has evolved into what is now one of north-west Europe's key economic and cultural regions. The Netherlands is a country with nearly 18 million inhabitants who lead their lives on what were once completely uninhabitable spaces, relaxing at the weekend, taking their children to swimming lessons. This is what is special about the Netherlands, a country where rivers have been tamed and whole seas transformed into fertile farmland. In the nineteenth century the water tradition laid the foundation for a prosperous state with a sophisticated administrative system.

After the Delta Works reached their milestone in 1986 with the completion of the Oosterscheldekering, a dam that, due to increasing environmental awareness, was turned into a storm surge barrier that could be opened and closed instead, the Dutch thought they were safe forever. Awareness of the dangers presented by the water declined and, driven on by the glory of the impressive works, Rijkswaterstaat (the Directorate General for Public Works and Water Management) and the water boards maintained the country's water

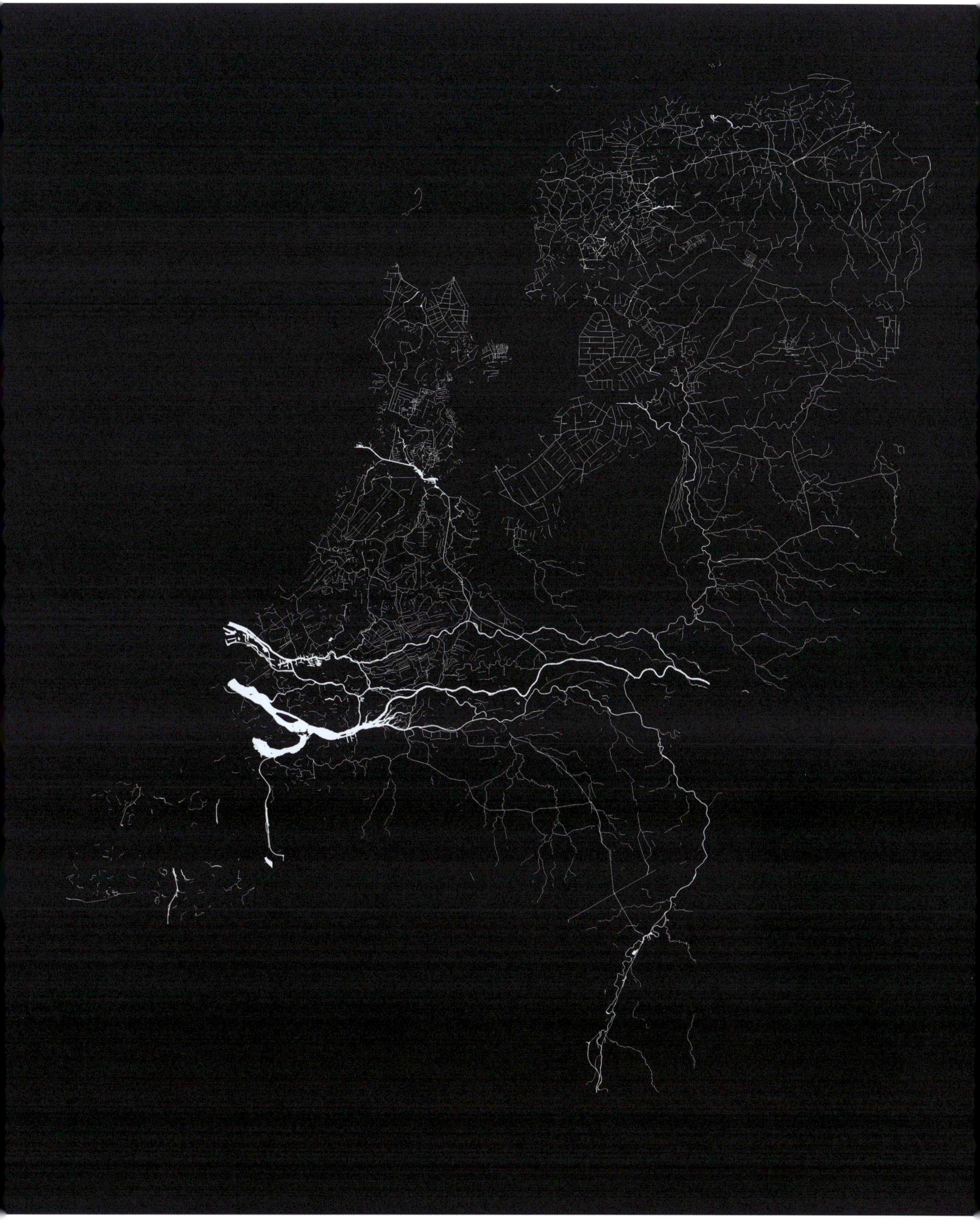

Hondsbossche and Pettemer Sea Defence – A trailing suction hopper starts constructing the sandy flood defence, 2014

infrastructure. This maintenance focused on reinforcing and raising weak spots in the river dikes – and this is where, from the 1970s, residents started protesting as orchards and historic buildings were demolished and dikes were turned into high, artificial embankments. The government responded by commissioning studies of alternative solutions, resulting in a design revolution led by a vanguard at Rijkswaterstaat and Staatsbosbeheer (the state forestry commission). Instead of making dikes higher and keeping rivers in their place it should be possible to give those very rivers room to spread out. The rivers' natural floodplains could be used to improve their discharge capacity instead of barricading villages and cities behind high basalt-clad embankments. Although Minister of Transport, Public Works and Water Management Neelie Kroes was certainly interested, the government had to cut spending and the necessary financing was not forthcoming. The new approach was embraced by Rijkswaterstaat teams who had worked for the organization's now defunct Deltadienst, which had overseen the Delta Works, but the Dutch parliament was not yet ready for such a drastic change in thinking.

Forty years after the 1953 disaster the country was rudely awakened from its slumbering awareness of the dangers presented by the water. This time the danger came not from the sea but from the hinterland. The extremely high water levels in the Rhine, Meuse and Waal rivers in January 1993 and again in 1995 made the dike authorities question the dikes' reliability. In 1995, shortly before part of the Waaldijk in Ochten broke – albeit on the safe river side – 250,000 people in Gelderland were evacuated. The Netherlands was briefly world news. Although the concept of climate change had not gained common currency yet, flood protection was suddenly high on the political agenda. From 1995 it was all hands on deck. The Deltaplan Grote Rivieren (Delta Plan Major Rivers), Ruimte voor de Rivier (Room for the Rivers), and the Maas works have been virtually completed. Growing expertise about flood defence and climate change over the last ten years has led to renewed inspections and reinforcements of the national dike system. New flood defence norms have been developed and new types of collaboration trialled. Until 2050 more than 1,900 kilometres of dike will be improved. Hydraulic engineers are implementing these projects with dredger companies, ecologists, urban planners and (landscape) architects to develop the new Dutch delta. This collaboration is what makes Dutch water management unique.

The social upheaval caused by flooding was made even clearer in August 2005, when Hurricane Katrina struck New Orleans, killing 1,200 people and causing enormous economic damage. Shortly afterwards, our awareness of climate change was given another jolt by Al Gore's film *An Inconvenient Truth*. Both events led the government to set up the Taskforce Management Overstromingen (TMO, a flood management task force) in November 2006. Its remit: by the end of 2008 the Netherlands must be organizationally better prepared for flooding, with a key emphasis on limiting human suffering and social dislocation. A specially appointed Delta Committee chaired by former government minister Cees Veerman published its recommendations in 2008, upon which the Delta Programme was launched in 2010 under the leadership of the independent Delta Commissioner, Wim Kuijken.

The course followed in recent decades has been characterized by a remarkable transformation of the traditional struggle *against* the water into an approach that works *with* the water. The force, will and qualities of the water itself are central – the water that, just like human beings, has its own identity and character. We must give this water room: to drain the rain and meltwater from the Alps, reduce flooding due to increasingly heavy downpours, and to cope with rising sea levels. This has led to innovative solutions such as the 'sand motor' along the coast, secondary channels in rivers and dual spatial use of the Waddendijk on the island of Ameland. Also, it has become clear that flood protection measures can be 'linked' to spatial planning projects: enhancing the appearance of a village, for example, or promoting environmental recovery that leads to increased biodiversity. This 'linkage' of spatial projects from other fields is not new – consider the Delta Works – but the fact that it is no longer initiated exclusively by the central government is. The collaboration between national, provincial and local governments, water boards, consultancies and dredgers as well as the residents of areas where water plays a role is a fine-tuning operation whose impact and clout are comparable with the discovery of the polder windmill. The scale and impact of the works form a series of interventions that will last until 2050. By then, every individual behind dikes or dunes will enjoy a the same high level of protection, with a risk of dying as a result of flooding not exceeding 1:100,000 per year. In places where the consequences would be very severe in terms of victims or economic damage, the level of protection will be even higher.

The Dutch delta is back on the drawing board for the umpteenth time in its existence, and the results are visible in the coastal and river landscapes. This laboratory of landscape art draws together the interests of agriculture and nature, of industry and ecology, and of recreation and shipping. The Netherlands will not only become safer but also more varied and attractive. The results form the basis for the projects that are currently being implemented, with the year 2050 in mind. This book will explore thirty interventions along rivers and the coast that unite hydraulic engineering, cultural history, nature and human use in matchless water landscapes that will entice readers to embark on expeditions themselves.

Management and Control

Management of the Netherlands' water system is spread over three administrative tiers: central government, the provinces, and the water authorities. Rijkswaterstaat has an executive role, implementing projects and policies for the main water system (the sea, large lakes and major rivers); the water authorities do this for the regional water systems, including storage basins and polders. The provinces play a role in spatial planning and directing regional organization. Matters such as housing, traffic and communicating with residents are the domain of the municipalities.

All three administrative tiers are represented in Security Regions, which play a role in disaster management if a calamity is imminent.

The country is protected against water from the sea, the major rivers and lakes by primary flood defences. The water authorities are responsible for managing most of these flood protection systems. Rijkswaterstaat manages the major storm surge barriers and dams.

The Netherlands, a Land of Water

Water Authorities

The Large-Scale Programmes

Large-scale programmes such as the Delta Plan and the Delta Plan Major Rivers were implemented in response to the catastrophic 1953 North Sea flood and the extreme high water levels in 1993 and 1995. Flood defences had to be upgraded as quickly as possible.

The high water levels in the 1990s led to plans that included a new approach to dealing with water. Instead of responding to emergencies after the event, the focus would now be on dealing with future problems before they had even occurred. There was a growing realization that the water must be given more room. Large-scale programmes such as the Maaswerken (Meuse Works) and Room for the River were established to implement this new thinking. Legislation was also passed requiring the periodic assessment of the primary flood defences to see whether they meet legally established norms. This way, any reinforcement that is required can be implemented before a problem arises.

The central government and the water authorities set up flood protection programmes to ensure that flood defences that did not conform to the norms during the assessment were upgraded. A new flood protection programme focuses on reinforcing the flood defences that did not conform with the norms during the most recent assessment. This ongoing programme also deals with flood defences that have to be reinforced so that they meet the new norms for 2050.

The large-scale programmes are all part of the Delta Programme, in which the central government, water authorities, provinces and municipalities all work together, using input from social organizations, business, citizens and knowledge institutes.

Time span

1990 2000 2010 2020 2030 2040

Budget 944 million euros (price level 2016)

Delta Plan Major Rivers

Reason: **high water in 1993 and 1995**
Aim: **implement urgent dike reinforcements on approx. 1,000 km of dike and embankment trajectories along the major rivers and dikes along the IJsselmeer and in the south-west delta. The measures along the Meuse were later continued under the Maaswerken (Meuse Works) programme**

— number of projects 214

Time span

1990 2000 2010 2020 2030 2040

Budget 558 million euros

Maaswerken (Meuse Works) (Grens- en Zandmaas)

Cause: **high water in 1993 and 1995**
Aim: **flood protection Meuse, improve spatial quality and self-funding through raw material extraction**

○ number of projects 55

Time span

1990 2000 2010 2020 2030 2040

Budget 740 million euros

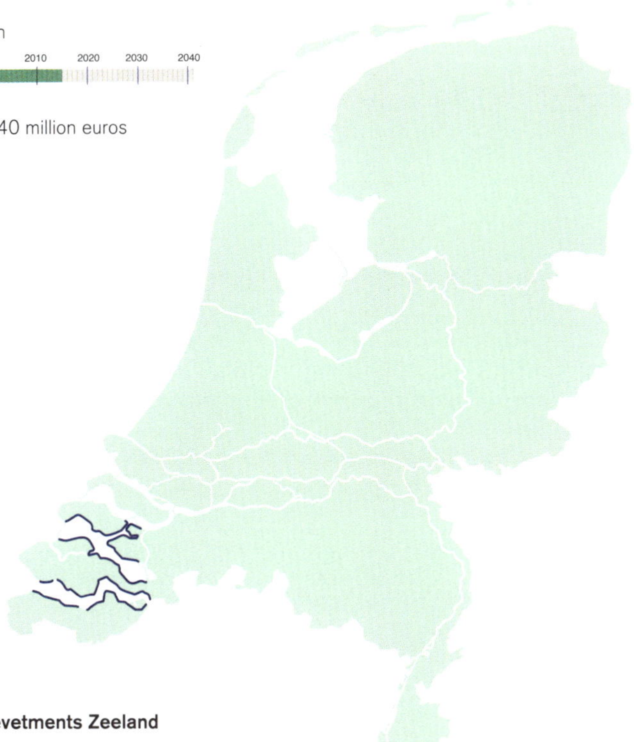

Stone revetments Zeeland

Reason: **inspection showed that the stone revetments on the dikes along the Ooster- and Westerschelde were not strong enough**
Aim: **reinforce 325 km of stone revetments**

— number of projects 108

Time span

1990 2000 2010 2020 2030 2040

Budget 2,4 billion euros

Room for the River

Reason: **high water in 1993 and 1995**
Aim: **improve flood protection (from 15,000 m³/sec to 16,000 m³/sec water discharge) and improve spatial quality**

○ number of projects 34

Time span

1990 2000 2010 2020 2030 2040

Budget 2,7 billion euros

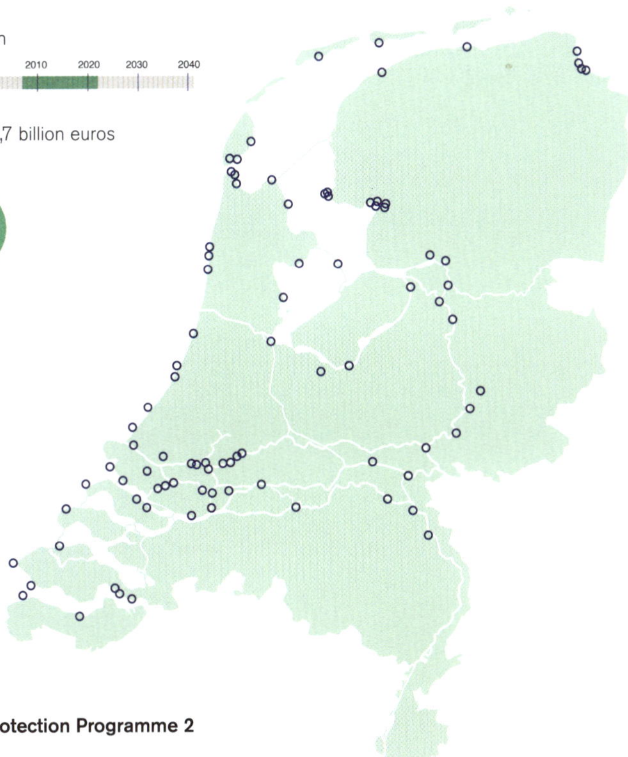

Flood Protection Programme 2

Reason: **Flood defences during the first (1996) and second (2006) assessments did not conform to legally prescribed norms**
Aim: **Reinforcing 362 km dikes, dams and dunes and 18 civil/hydraulic engineering works so they conform to the norms**

○ number of projects 87

Time span

1990 2000 2010 2020 2030 2040

Budget 4,9 billion euros

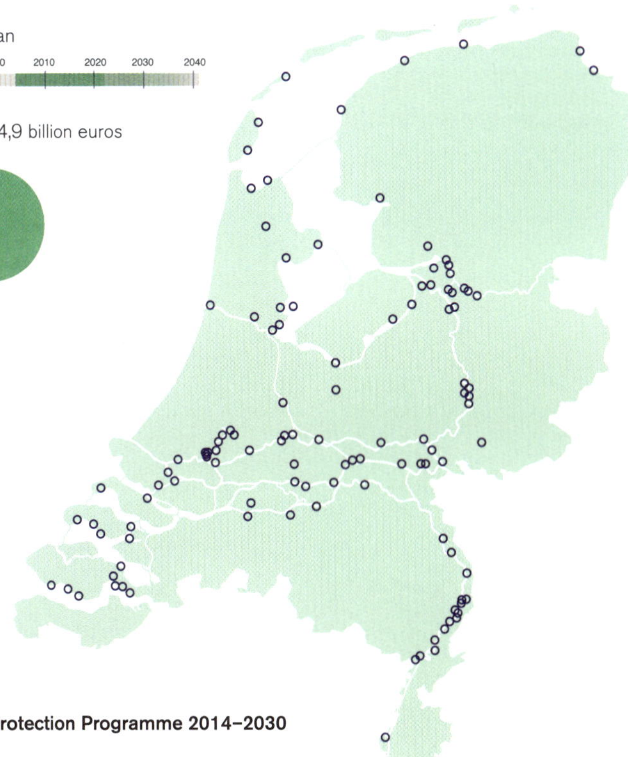

Flood Protection Programme 2014–2030

Reason: **Flood defences during the first (1996) and second (2006) assessments did not conform to legally prescribed norms**
Aim: **Reinforcing 362 km dikes, dams and dunes and 18 civil/hydraulic engineering works so they conform to the norms**

Tackling 803 km of dike and 468 civil/hydraulic engineering works has been programmed up to 2022. Some 140 km of dike have been added to this that have to be upgraded before 2050 to conform to the norms applicable as of 1 January 2017.

○ number of projects >100

New Flood Defence Policy

Since 1 January 2017 new flood defence norms have been in force for the Netherlands. These norms are not only based on the chances of flooding but also on the effects of flooding on particular areas and their inhabitants. The worse the effects, the stricter the norm. The *basisbeschermingsniveau* (basic protection level) applies to everyone in the Netherlands: the chance that someone will die as a result of flooding may not be greater than 1:100,000 per year. A region is given additional protection if a flood would result in widespread social disruption, with many dead and injured and considerable economic damage. The calculation takes into account the region's forecast situation in 2050 — its predicted population and economic value. The effects of a flood depend on circumstances behind the dike, the speed of flooding and the water level.

The 2017 Wettelijk Beoordelings-instrumentarium (WBI) is a new system for assessing whether the norms are being met. The first round of assessments will take place between 2017 and 2023. This will provide a nationwide overview that will be reported to the Dutch parliament in 2023.

FAILURE MECHANISMS

A Overflowing and wave overtopping
The water level is higher than the crest, resulting in flooding. The flood defence can then collapse as a result of continuous erosion.

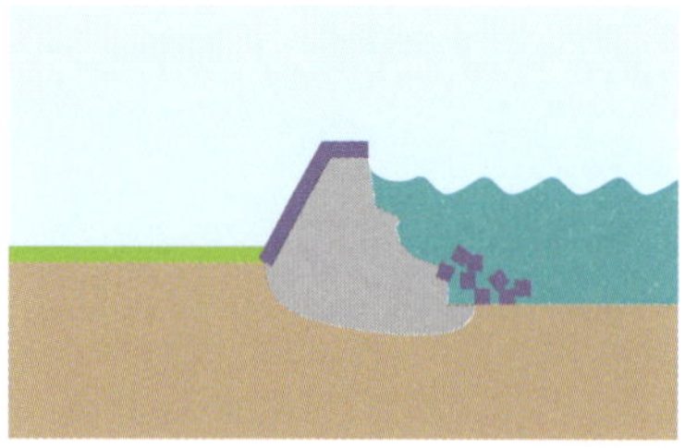

B Unstable revetment
Waves can also damage the slope. Clumps of grass can be swept away from turf. This can also happen with the stone blocks used in hard revetments.

C Macro-instability inner slope
Water pressure on the flood defence and increased substrate water tension or infiltration of floodwater can cause the inner slope to slide off.

D Macro-instability outer slope
The outer slope can slide off if the outer water level drops rapidly after high water.

E Piping
Piping as a result of seepage flowing through the substrate, carrying soil with it and undermining the dike.

F Unstable foreshore
Erosion of the outer slope by waves or current.

G Micro-instability
Instability of the inner or outer slope caused by seepage flowing out of the earth body.

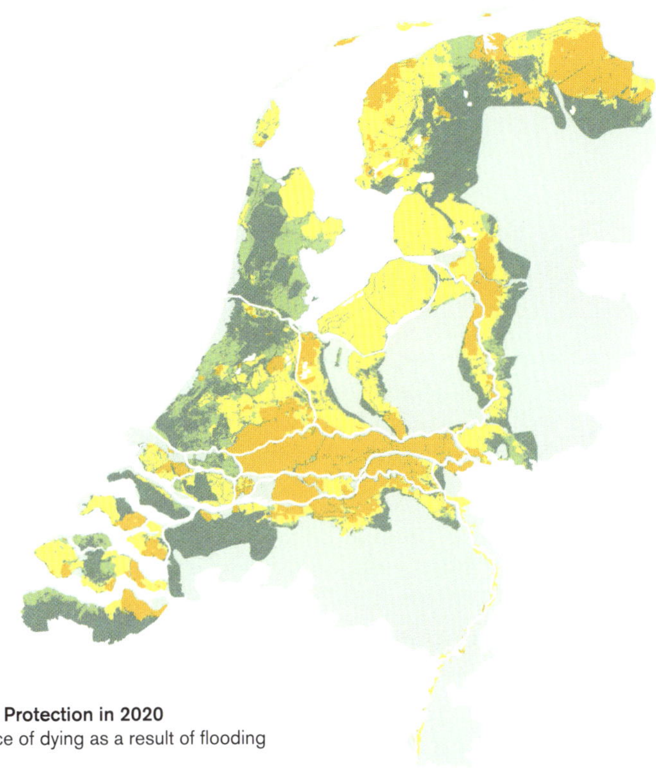

Flood Protection in 2020
Chance of dying as a result of flooding

Meets future *basisbescherming* (basic protection level)
- smaller than 1:1,000,000
- between 1:1,000,000 and 1:100,000

Does not meet *future basisbescherming* (basic protection level)
- greater than 1:100,000
- area potentially at risk of flooding

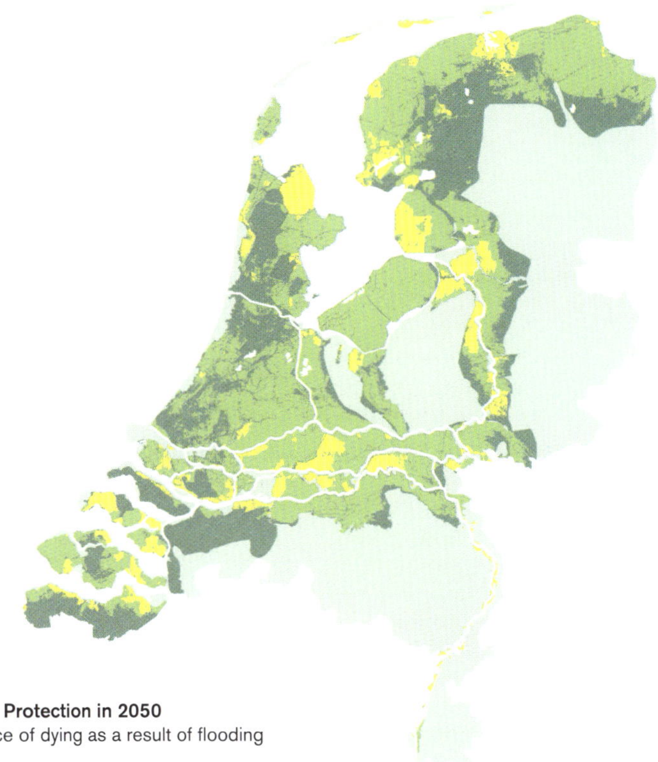

Flood Protection in 2050
Chance of dying as a result of flooding

Meets future *basisbescherming* (basic protection level)
- smaller than 1:1,000,000
- between 1:1,000,000 and 1:100,000
- area potentially at risk of flooding

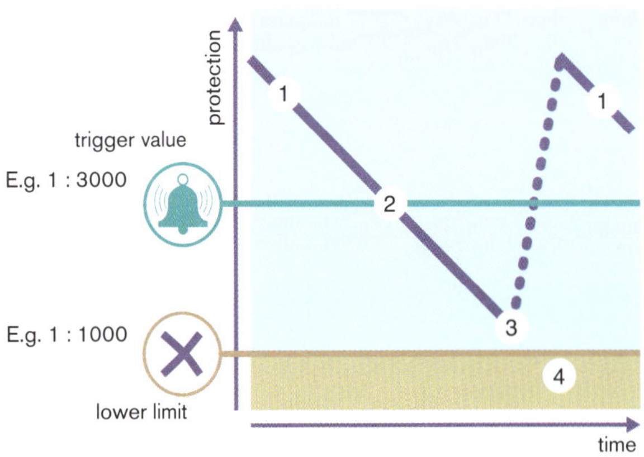

The course of flood protection during the lifespan of a dike
1 Protection level decreases due to higher water levels (as a result of climate change) and declining dike strength (as the dike ages).
2 Preparations can start as soon as the trigger value is reached.
3 The dike is reinforced before the lower limit is reached.
4 Under the lower limit the dike will not comply with the prescribed norms.

Assessing Flood Defences

Primary flood defences are periodically assessed in accordance with hydraulic engineering criteria. The 2009 Water Act stipulates two values that together form the norm for every dike trajectory: a trigger value and a lower limit. The trigger value indicates that timely preparations must be made to reinforce the dike. The lower limit indicates the minimum requirements the dike must meet; it is not allowed to fall below this level because it will not be able to provide the prescribed protection.

If a flood defence does not comply with the trigger value, measures will be taken to reinforce it. There is at least ten years time to implement this process, because the trigger value is stricter than the lower limit. The timing of the actual reinforcement depends on various factors, e.g. urgency, the availability of financing, the complexity of the measures and the capacity to implement the measures.

Use these tabs to match interventions with projects
01 02 03 04 05 06 07 08 09 10 11 12 13 14 15 16 17 18 19 20 21 22 23 24 25 26 27 28 29 30 31 32 33

Sample Sheet of Interventions

Dutch rivers have a characteristic cross-section, with a narrow summer bed and a wide winter bed. For most of the year, the river flows between the summer dikes. This is the summer bed, the navigation channel. When the water level is high, the river flows over the summer dikes and inundates the floodplains. These flood defence areas are, in turn, protected by winter dikes. The course of the rivers can be adapted in various ways to increase their discharge capacity and improve flood protection. A distinction is sometimes made between dikes and embankments: water is allowed to flow over a embankments but this must never happen with a dike because it is meant to protect the area directly behind it. Applying this logic, it is more accurate to talk of summer embankments, and winter dikes, but the distinction is not always clear cut. The Meuse embankments in Limburg, for example, are primary flood defences and therefore function as dikes. This is also true of the quays along regional waters.
Along the coast, lakes and sea arms, protection against the impact of waves is of the utmost importance. Flood defences here are therefore often reinforced with hard materials. Half of the height of these dikes might be necessary to absorb wave uprush. This uprush can be reduced by putting foreshores, dams, berms and vegetation in place.

STRONGER DIKES 01-07

01 Reinforcing Dike Revetment

A dike can be reinforced by improving the revetment used for the dike body. A great deal can be achieved by simply applying a new top layer of clay with high-quality turf (see p. 78, 182). Besides this, a wide assortment of hardening materials is available that can withstand the impact of waves: concrete blocks and columns, rockfill, rubble or open-cell concrete blocks, which may or may not be combined with liquid asphalt (see p. 62, 70). Examples also exist of stones being glued to the existing dike body. Fitting the dike with wave attenuators or wave absorbers provides a (partial) alternative to dike raising (see p. 66, 70).

02 Dike Widening

Widening a dike makes it more robust. An additional advantage is that the dike can fulfil other functions than only flood protection. It can be reinforced by adding a shoulder berm or adjusting the entire slope, which is possible for both the inner and outer side of the dike (see p. 196).) A gentler slope creates more room for other functions such as a cycling path on the slope of the dike (see p. 70).

03 Dike Raising

The traditional solution to flooding is to make dikes higher. This can be done using soil and clay but alternatives include masonry, concrete or some other material (see p. 70, 90, 182, 192, 196). Because the slope of a dike must not be too steep, the dike is also widened if it is raised in this way.

04 Internal Dike Reinforcement

Besides adjusting the dike's shape, numerous options to strengthen it internally also exist. Diaphragm, pile or sheet pile walls are often used, installed using vibration-free techniques to prevent damage to existing buildings (see p. 74). Dike nailing holds the dike mass together, comparable to reinforcing concrete. Rows of nails, each several metres long, are drilled into the dike. This was done at the Lekdijk in Vianen's historic town centre. Dike dowels are steel tubes that are stuck into the dike like giant skewers. Filled with grout, a sort of cement, they keep the dike anchored.

05 Raising Quays

Raising a dike or embankment by installing a wall or partition on it. This structure can be incorporated into the urban design so that the flood defence is integrated into buildings or steps (see p. 192).

06 Seepage Screen

High water levels can cause water to seep under a dike and wash out sand. This is known as 'piping' and it can threaten the stability of the dike. Seepage screens can be installed up to 20 metres deep to prevent seepage in the dike and sand being washed away. The screens are made of geotextile (a type of cloth or canvas), cement bentonite, which can be poured into deep grooves, or sheet pile walls that are vibrated or pushed into the ground (see p.78, 120, 236)

(NEARLY) INVISIBLE REINFORCEMENT 08-09

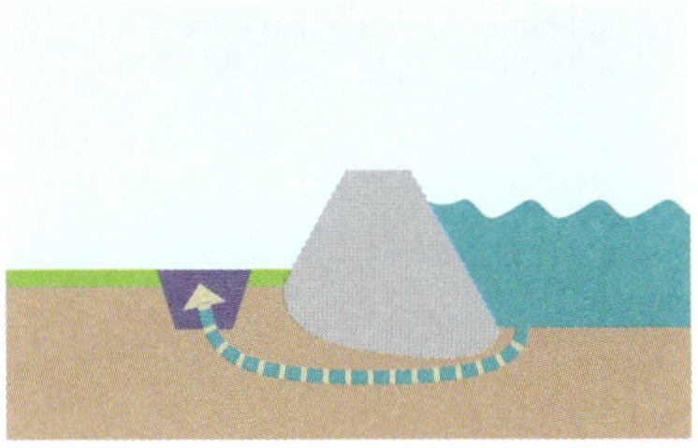

07 Drainage Techniques
When water levels are high, the weight of the water can cause groundwater to dam up. This then suddenly spouts out of the ground (far) behind the dike, out of a drain, cellar or ditch. To reduce the pressure of the groundwater a ditch ('seepage catchment'), drains or coarse sand filters are constructed behind the dike. The seepage water can accumulate here and be pumped away (see p. 84, 120).

08 Removable Flood Defence
Raising dikes in villages and cities in particular is often impossible or undesirable. The classic solution in an emergency is to build a temporary flood defence. This is still effective and can be perfected technically by using light materials (aluminium) and waterproof joints (profiles, rubber strips). The best solution can be selected for each opening – stop logs, lifting gates, flap gates, flood defence doors to name just a few (see p. 200. Constructing a temporary flood defence is labour intensive and requires considerable practice.

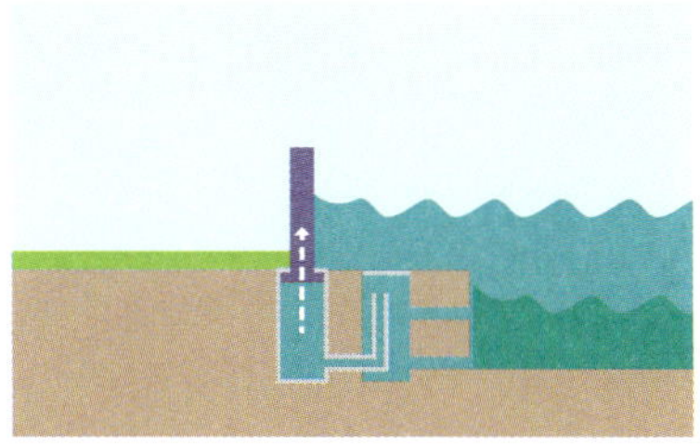

09 Floating Flood Defence
The self-raising flood barrier is a special collapsible flood protection measure. It is concealed in the quay and automatically floats under the pressure of rising water (see p. 206). This measure makes a permanent flood defence unnecessary, so that the historic cityscape and the experience of the water remain unaffected.

MORE ROOM FOR WATER 10-18

10 Dike Repositioning
A river's capacity to cope with water is increased not only by a higher dike but also by a wider bed. Dike repositioning can therefore be an alternative or complementary to dike raising. Shifting the winter dike inland widens the floodplains and gives the river more space (see p.136, 188, 236, 248).

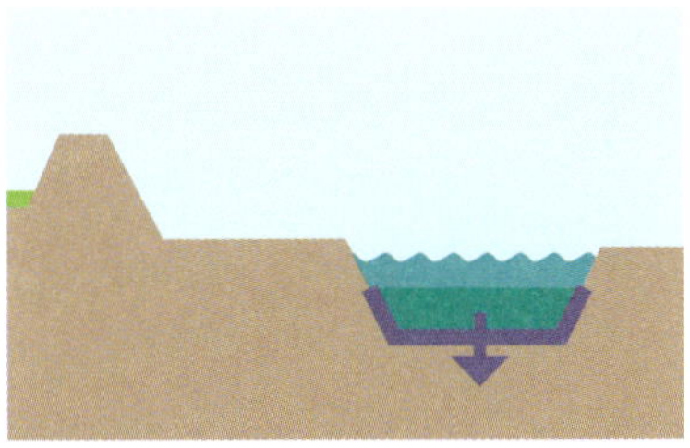

11 Deepening the Summer Bed
The river stays within its summer bed when the water level is normal or low. Deepening this channel increases the discharge capacity and slows the rate at which floodplains are inundated. This intervention has not been included in the book.

12 Lowering Perpendicular Groynes
It is important that rivers remain navigable during dry spells. In the past this was guaranteed in many places by installing groynes in the summer bed. These short dams are built perpendicular to the river. They prevent the water meandering and the river silting up. But as a result of erosion, the groynes are now too high, damming up the water unnecessarily in the case of high water. By lowering groynes an optimum can be sought, which is happening in the Waal river. An alternative solution to retain navigability comes from Germany and has been implemented in an adapted form to this river: constructing kilometres-long longitudinal dams with a main channel for rapid navigation and side channels for nature to develop. This intervention has not been included in the book.

13 Removing Obstacles
In the floodplains all sorts of obstacles can be found that prevent a rapid discharge of river water: vegetation, buildings (brickworks, farms), or ferry access dams. Removing these obstacles improves the water flow when water levels are high (see p. 116).

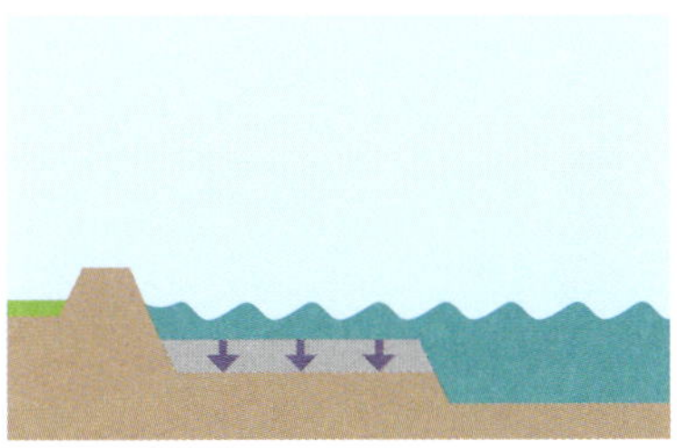

14 Lowering the Floodplain
The floodplains are the areas where rivers are allowed to flow over their banks. A layer of sludge is left after each flood, eventually leading to thick accumulations of clay. Lowering the floodplains creates more room for high water. When all the clay that has accumulated over the centuries is removed, the old sandy bed reappears on the surface, including the contours of sand ridges and channels. This is known as 'contour-guided clay extraction' (see p. 168, 230).

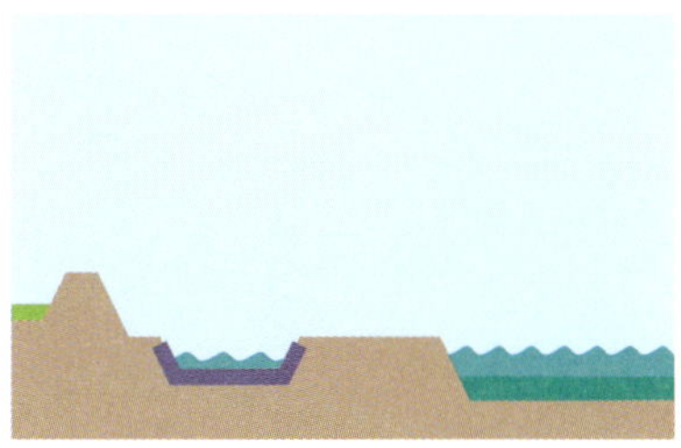

15 Secondary Channel
Besides the lowering of floodplains, digging secondary channels in the winter bed can also create more room for the river. When the water level is low the water in these channels is stagnant and fulfils an important ecological function. When the water level is high these channels fill up and join the main flow of the river and thus increase the discharge. Nature development alongside rivers is often accompanied by the construction of secondary channels (see p. 136, 168, 226, 230, 236, 242, 248).

MORE ROOM FOR WATER

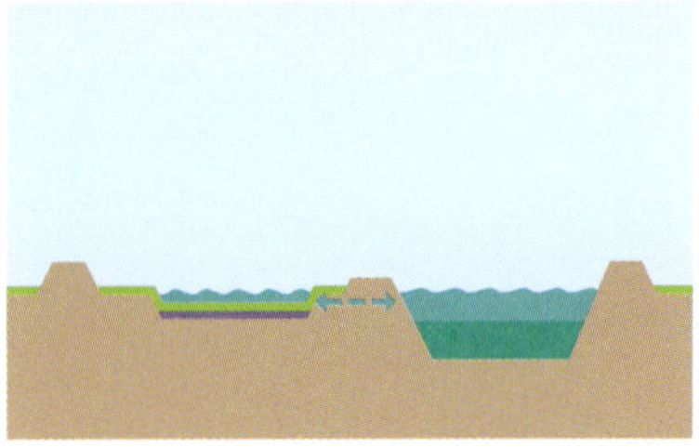

16 High water Channel
If a secondary channel is mostly dry and only fills up with water when the water level is high, it is called a high water channel. A 'green river' like this is a diked part of the flood-plains but it can also flow in an alternative route around the floodplains. Some secondary channels are flooded annually, for others the chance of this happening is 1 in 80, 100 or 250 per year (see p. 126, 140). These channels can be used for nature, recreation and/or agriculture for the rest of the time.

17 Bypass
Side or high water channels are regarded as bypasses when they direct the water for long distances around a narrowing in a river. This happened at Reevediep, a bypass of the IJssel river south of Kampen. This intervention has not been included in the book.

18 Flood-free Areas
Flood-free areas such as raised shelters or *terpen* (artificial mounds) can be built in flood zones such as floodplains or polders that are designated water storage areas. These elevated safe areas are suitable for housing and other buildings (see p. 140). An alternative is to construct a ring dike around existing urban settlements in order to protect the buildings from flood damage (see p. 160).

WATER STORAGE 19-20

19 Depoldering
Making a polder partly or completely subject to flooding is called depoldering: when water levels are high the polder is turned into a water storage area. This function can be combined with the existing agriculture but it also links up well with nature development and/or recreation (see p. 132, 136, 140).

20 Water Storage Area
A severe storm sometimes coincides with an exceptionally high river discharge. At times like this the water cannot flow out to sea because the sea level is too high (IJsselmeer) or because the storm surge barriers are closed (the Zeeland and Rhine-Meuse deltas). Vast amounts of water then need to be stored for a short period of time. Water storage areas such as the Volkerak-Zoommeer in Zeeland are designed for this purpose. This intervention has not been included in the book.

REINFORCING THE COAST 21-24

21 Beach Extension
The threat of high water from the sea can be reduced by extending the beach towards the sea. This reinforces the coast line. The extended beach breaks the waves and provides room for the sea flood defence to be widened through sand nourishment. At some distance from the coast a dredging ship sucks sand from the seabed and spouts it up near the coast (see p. 94, 100).

22 Dune Reinforcement
The coast can be reinforced through an upgrade of the dunes: widening, raising or reinforcing existing dunes with sand to make them more robus. This intervention has not been included in the book.

23 Outer-dike Dune Construction
Instead of raising sea dikes, a new row of dunes and a beach can be constructed in front of the dike. This approach links flood defence and recreational use (see p. 220).

24 Inner-dike Dune Construction
In places where there is not enough room next to a dike for reinforcement or dune construction, it is possible to construct a new row of dunes behind the old existing embankment. The outer-dike area between the and the new flood defence is ideal for saline farming and nature development (see p. 84).

VARIATIONS ON A DIKE 25-33

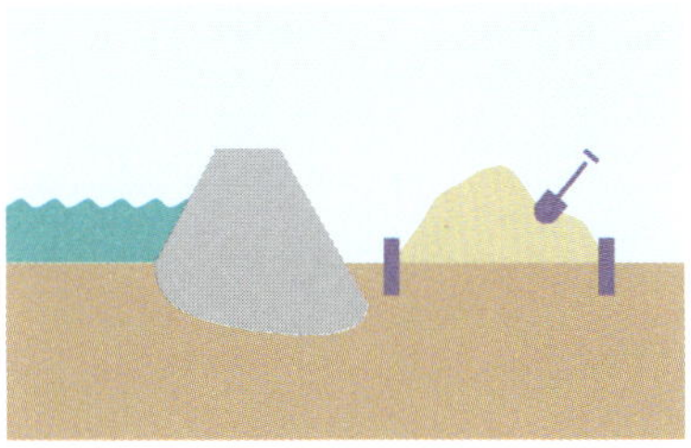

25 Earthquake-proof Dike
Groningen is prone to earthquakes because of the gas extraction that takes place there. An earthquake can cause a sort of quicksand to form underneath a dike, known as liquefaction. If this happens, the dike might partly collapse. The principle of an earthquake-proof dike is to make it extra high or wide, allowing it to collapse slightly rather than collapsing altogether in case of a strong earthquake. A supply of clay is stored in a nearby depot so that the dike can be repaired quickly. (see p. 9).

26 Dike-in-Dune
In Katwijk the dune was reinforced by building a dike (clay with a stone revetment) into it. The dike was covered with dune sand. This approach makes the raising of a dune unnecessary and maintains the sea view in bathing resorts (see p. 100). This combination of 'hard' and 'soft' structures is called a hybrid flood defence.

27 Double Dike (near the Sea)
Constructing a second dike behind the primary embankment to create a brackish ecological system in the intervening space with the use of culverts. This is ideal for nature development and saline aquaculture. During high water the culvert can be shut, thus closing off the flood defence (see p. 84, 90).

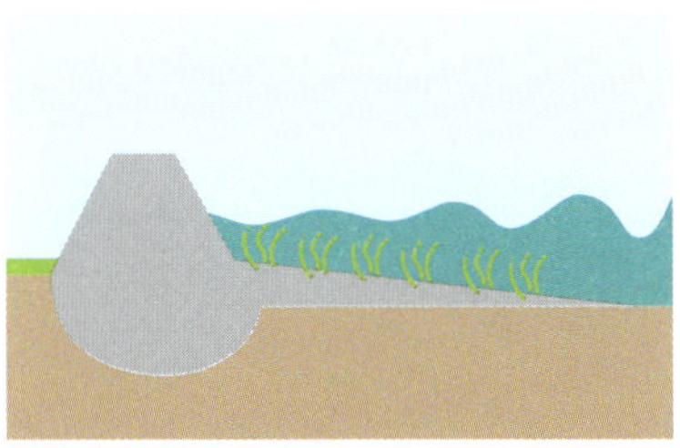

28 Wave-reducing Dike
Interventions to reduce the impact of waves can make the raising of a dike partly or completely unnecessary. This can be achieved by exploiting a dike's own capacity to attenuate wave uprush (a rough surface, built-in cavities), a technique that has been used in the sea defences (see p. 66, 70, 90). In the river areas this concept has been realized differently. There, wave-reducing vegetation as has been planted in front of dikes, like the stand of willows in the Noordwaard (see p. 140).

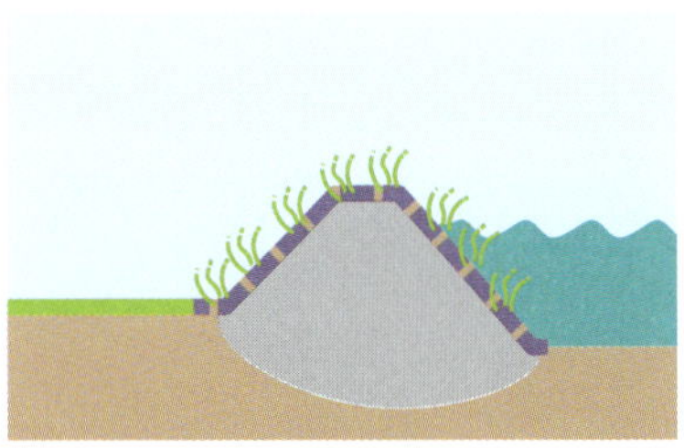

29 Rijke Dijk (Rich Dike)
Stone revetment (concrete, rubble) accommodated with cavities that act as wave absorbers and as habitats for flora and fauna (see p. 66, 70, 90). This creates a more natural transition between sea and dike. River dikes are sometimes sown with a mix of different seeds to enhance biodiversity.

30 Smart Dike
Dikes equipped sensors that make it possible to continuously measure their stability at a distance. Sensors register movements, temperature or humidity in the dike. When the dike is under pressure, during a storm, for example, any deviations from the norm can be noticed and acted upon quickly (see p. 90, 94).

31 Multifunctional Dike: Climate Dike or Delta Dike
Some dikes are made higher or stronger than the current standards require. Their oversize is expected to make future adjustments to climate change unnecessary for a long time. The reason for this might be to store excavated soil on the spot but also to create room on the dike for other functions such as a boulevard. The dike then becomes part of its surroundings, enhancing the quality of life in a region. Climate dikes are future-proof and multifunctional (see p. 74, 136). One variant is the climate dune (see p. 84)

32 Grandstand or Terrace Dike
A dike with terraces in the dike berm. The terraces break the height, reinforce the dike as a shoulder and can be used for recreation, vegetation or even housing. The multifunctional grandstand dike in the Brakelse Benedewaarden is called the Wakkere Dijk ('alert dike') (see p. 136).

33 Multifunctional Dike: Dike-in-Boulevard
Reinforcing and upgrading an existing boulevard along the beach using sand, sheet pile walls, geotextile, concrete columns, liquid asphalt, blast furnace slag and/or rubble (see p. 9).

Programmes 1995-2022

This map shows nearly all of the projects that have been implemented in the Netherlands in recent decades to improve flood protection. Most of the programmes referred to here have been virtually completed. The most recent Flood Protection Programme started in 2014 with the aim of upgrading more than 700 kilometres of primary water-retaining structures (dikes) and more than 200 flood-protection hydraulic works (locks, barrages) that currently fail to meet the required standards. These projects are also shown on the map.

The Flood Protection Programme is an ongoing programme. It is carried out jointly by Rijkswaterstaat and the water authorities implementing a succession of six-year programmes. The programme's aim is to ensure that all flood defences conform to the new norm by 2050.

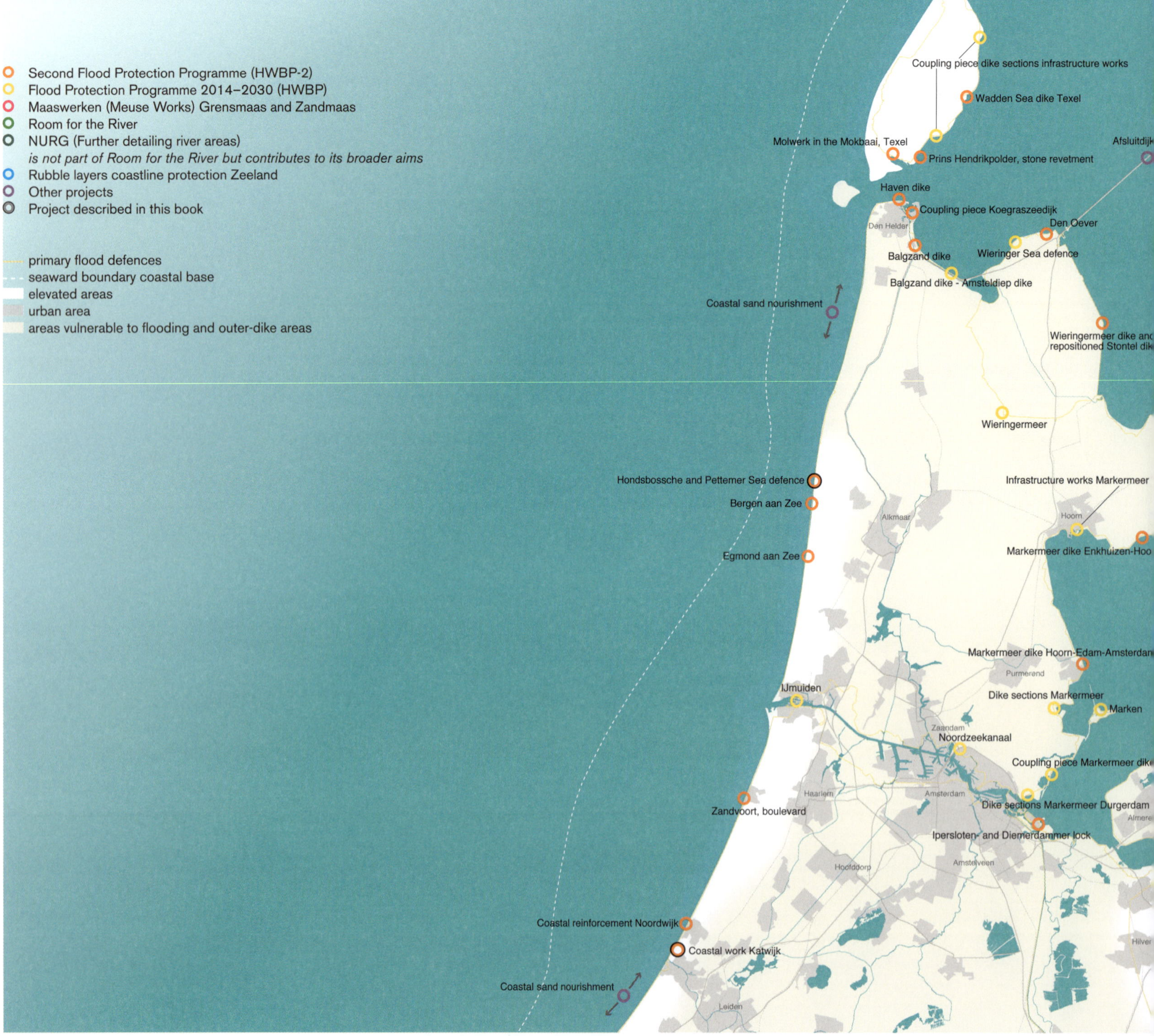

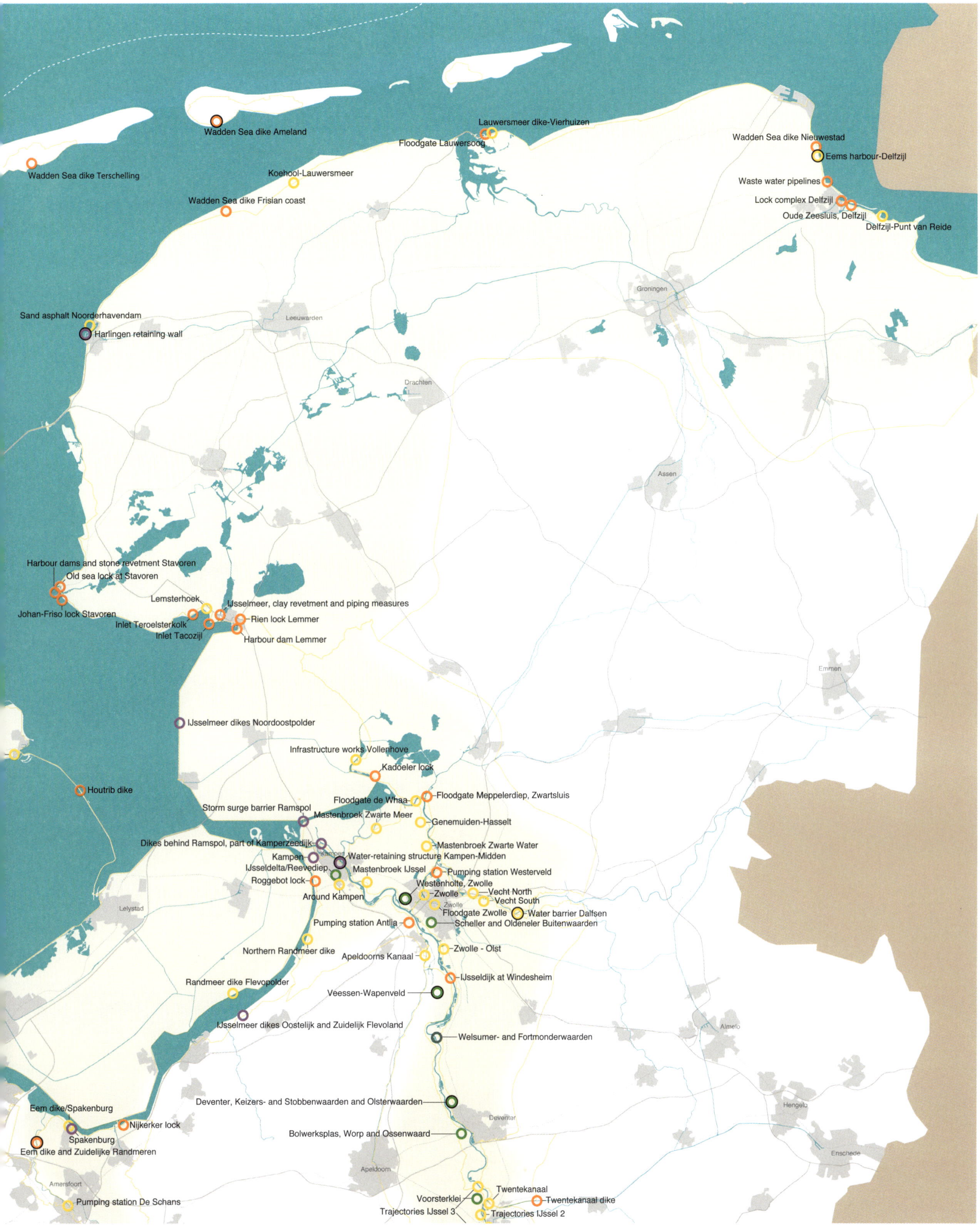

Wadden Sea dike Ameland
Lauwersmeer dike-Vierhuizen
Floodgate Lauwersoog
Wadden Sea dike Nieuwestad
Eems harbour-Delfzijl
Wadden Sea dike Terschelling
Koehool-Lauwersmeer
Waste water pipelines
Lock complex Delfzijl
Oude Zeesluis, Delfzijl
Delfzijl-Punt van Reide
Wadden Sea dike Frisian coast
Groningen
Sand asphalt Noorderhavendam
Harlingen retaining wall
Leeuwarden
Drachten
Assen
Harbour dams and stone revetment Stavoren
Old sea lock at Stavoren
Lemsterhoek
IJsselmeer, clay revetment and piping measures
Johan-Friso lock Stavoren
Rien lock Lemmer
Inlet Teroelsterkolk
Inlet Tacozijl
Harbour dam Lemmer
Emmen
IJsselmeer dikes Noordoostpolder
Infrastructure works Vollenhove
Kadoeler lock
Houtrib dike
Floodgate Meppelerdiep, Zwartsluis
Floodgate de Whaa
Storm surge barrier Ramspol
Mastenbroek Zwarte Meer
Genemuiden-Hasselt
Dikes behind Ramspol, part of Kamperzeedijk
Mastenbroek Zwarte Water
Kampen
Water-retaining structure Kampen-Midden
IJsseldelta/Reevediep
Mastenbroek IJssel
Pumping station Westerveld
Roggebot lock
Westenholte, Zwolle
Zwolle
Vecht North
Vecht South
Around Kampen
Lelystad
Floodgate Zwolle
Water barrier Dalfsen
Pumping station Antlia
Scheller and Oldeneler Buitenwaarden
Northern Randmeer dike
Zwolle - Olst
Apeldoorns Kanaal
IJsseldijk at Windesheim
Randmeer dike Flevopolder
Veessen-Wapenveld
IJsselmeer dikes Oostelijk and Zuidelijk Flevoland
Almelo
Welsumer- and Fortmonderwaarden
Eem dike/Spakenburg
Deventer, Keizers- and Stobbenwaarden and Olsterwaarden
Hengelo
Deventer
Nijkerker lock
Spakenburg
Bolwerksplas, Worp and Ossenwaard
Eem dike and Zuidelijke Randmeren
Enschede
Apeldoorn
Amersfoort
Twentekanaal
Voorsterklei
Twentekanaal dike
Pumping station De Schans
Trajectories IJssel 3
Trajectories IJssel 2

Boulevard Scheveningen
The Sand Motor
Delfland coast
Pumping station Westland
Maeslantkering
Hinterland Maeslantkering
Coast of Voorne
Storm surge barrier Hollandsche IJssel
Capelle a/d IJssel
Capelle/Moordrecht
Hille dike
Spuihaven Schiedam
Oude Maas/Voorne Putten
IJsselmonde North and South
Brielse Maasdijk
Hellevoetsluis
Flaauwe Werk
Flow slide V3T
Norm trajection no. 20-3
Spui East
Flow slide Spui
Spui West
Oude Maas/Hoeksche Waard
Buitenhavendijk Middelharnis
Northern beach Schouwen
Nieuwe Stadse Zeedijk
Schelphoek
Burghsluis-Schelphoek
Burghsluis
Flaauwershaven/Borrendamme
Coastal sand nourishment
Zierikzee
Inlet Zuidhoek/Bruinisse
Anna Frisopolder
Onrustpolder
Stone revetments Zeeland
Sint-Annaland
Volkerak-Zoommeer
Oost-Bevelandpolder (midden)
Wemeldinge
Wemeldinge Oost
Zuid-Beveland Oost, Oosterschelde
Westkapelle
Koopmanshaven lock
Vlissingen
Channel Zuid-Beveland
Zuid-Beveland West, Westerschelde - Borssele
Zuid-Beveland West, Westerschelde - Hansweert
Veerhaven, Kruiningen
Zuid-Beveland Oost, Westerschelde
Zimmermanpolder
Hoedekenskerke
Waarde- and Westveerpolder
Molenpolder (West)
Emanuelpolder
Kruiningenpolder
West Zeeuws Vlaanderen
Vlissingen
Port Scaldis
Borssele
Zuid-Beveland West, Ritthem - Baarland
Ellewoutsdijk
Waterdunen
Nieuw-Neuzenpolder (East)
Eendrachtspolder (East and West)
Kleine Huissenspolder (West)
Margarethapolder (West)
Lock complex, Terneuzen
IJssel dike Gouda
Stolwijkersluis
Lek, Alblasserwaard and de Vijfheerenlanden
Rest assignment Hollandsche IJssel
Strong IJssel dikes Krimpenerwaard
Capelle/Zuidplas
Nederlek
Krimpen
Krimpen a/d Lek
Ostmolen dike
Streefkerk-Ameide
Bergambacht-Ammerstol-Schoonhoven
Dike improvement Kinderdijk - Schoonhovenseveer
Nederlekdijk
Papendrecht Ketelweg
Sliedrecht (Kleindiep and Baanhoek)
Hoeksche Waard North
Flow slide Dordtsche Kil
Island of Dordrecht East
Island of Dordrecht West
Hoeksche Waard South
Depoldering Noordwaard
Noordwaard
Avelingen
Merwede dike
Gorinchem-Waardenburg
Gameren
Brakelse Benedenwaarden and Buitenpolder Munnikenland
Jannezand
Steurgat/Land van Altena
High-water release Dongemond area
Amer/Donge
Geertruidenberg/Amertak
Bergsche Maas/Land van Altena
River widening Overdiepsche Polder
Drongelens kanaal
Heusden
Central Holland (C-barriers)
Hagestein-Everdingen
Honswijkerwaarden, barrage island Hagestein, Hagesteinse Uiterwaard and Heerenwaard
Hagestein-Everdingen at Vianen
Room for the Lek
Bank erosion Klaphek
Zederik
Vianen
Ameide Sluis-Fort Everdingen
Diefdijk
Beneden-Wa
Alphen a/d Rijn
Den Haag
Zoetermeer
Gouda
Delft
Utrecht
Nieuwegein
Rotterdam
Schiedam
Vlaardingen
Spijkenisse
Dordrecht
Breda
Tilburg
Roosendaal
Bergen op Zoom
Middelburg

Zutphen
IJssel pavilion, Vispoorthaven, Havenstraat
Cortenoever
Den Elterweg, Zutphen
Excavation of lower floodplains Renkumse and Wageninger benedenwaard and ferry slip Lexkesveer
Lower Rhine, Betuwe, Tieler- and Culemborgerwaard
Ede
Lower Rhine, Arnhemse- en Velpsebroek
De Tollewaard
IJsseldijk stadsgracht, Doesburg
Machinists' school Elst
Doorwerthsche Waarden
Amerongen-Beatrix lock
Rijnkade
Arnhem
Grebbe dike
Westervoortsedijk II
Industrial area Grutbroek
Pleijweg/Schaapdijk/Broekdijk, Arnhem
Meinerswijk
Spillways at Pannerden and Hondsbroeksche Pleij
Lek, Betuwe, Tieler- and Culemborgerwaard
Huissensche Waarden
Pannerden/Loo
Neder Betuwe
Middelwaard
piping measures
Tiel-A50
Sprok-Wolferen
A50 - Doornenburg
Municipality of Tiel
Afferdensche and Deestsche waarden
Lingewaard
Rijnwaardense Uiterwaarden
Heesseltsche uiterwaard
Midden-Waal
Lower city Nijmegen
Nijmegen
Tiel-Waardenburg
Dike repositioning Lent
Bemmelsche Waarden
Millingerwaard
Excavation of lower floodplains Millingerwaard
Ravenstein-Lith
Fort Sint Andries
Keent and Keent-Grave
Hemelrijksche Waard
Oss
Embankment Mook (Ottersum)
Deepening reach Lith
Floodgate Engelen
Adjustment Grave barrage
Quay population centre Mook & Middelaar
's-Hertogenbosch
Increasing groundwater level Grave
Quay population centre Gennep
Deepening reach Grave
Infrastructure works Boxmeer-Grave
Adjustment Sambeek barrage
Embankment Afferden
Boxmeer
Increasing groundwater level Sambeek
Nieuw-Bergen
Embankment Eiland van Bergen
Eroding banks pilot project
High-water channel Noord Well-Aijen
Well
High-water channel Zuid Well-Aijen
Ooijen-Wanssum
Helmond
Eindhoven
Arcen
Embankment Lottum
High-water channel Lomm
Deepening reach Sambeek (Venlo-Arcen)
Embankment Grubbenvorst
Venlo-Velden
Blerick Groot Boller
Inner city quai Venlo
Blerick
Embankment Venlo
Venlo
Baarlo
Steyl-Maashoek
Belfeld
Kessel
Beesel
Embankment Neer
Restoration Rijkelse Bemden
River widening Swalmen-Beesel
Buggenum
Retention area northern part of Lateraalkanaal-West
Quay population centre Roermond
Retention area Lateraalkanaal-West (extension)
Alexanderhaven
Retention area southern part of Lateraalkanaal-West
Barriers Roer
Heel
Embankment Merum
Thornwessem
Embankment Brachterbeek
Embankment Ohe en Laak
Embankment Aasterberg
River widening Roosteren
Embankment Roosteren
Secondary channel Visserweert
Grensmaas Koeweide
River widening Nattenhove
Waterway widening Grevenbicht
Soil and bank protection Berg aan de Maas
Sittard
Embankment Urmond
River widening Urmond
Embankment Maasband
Secondary channel Maasband
Embankment Meers
Geleen
Soil and bank protection Meers
River widening Meers
Embankment Geulle aan de Maas
River widening Aan de Maas
River widening Itteren
Soil and bank protection Itteren
River widening Borgharen
Heerlen
Embankment Maastricht West
Maastricht
Lock Bosscherveld
Water park Bosscherveld
Embankment Maastricht Oost
Embankment Eijsden

Depoldering Noordwaard – The water level rises, the residents are high and dry on the terp, 2016

The Nederlands' Second Nature

A New Layer in an Ancient Water Landscape

What follows is a random selection of news about the climate in the year of writing, 2017. In January Europeans found themselves in an upside-down world: in Scandinavia it was exceptionally mild, in southern Europe unusually cold. In Vienna people skated on the Danube but in central Italy an avalanche demolished an entire hotel. In June dozens of people were killed in Portugal by forest fires caused by drought. On 11 July half of the Netherlands was flooded following downpours that brought 50 millimetres of rain in many places. The next day an ice-shelf larger than the province of North Brabant broke away in Antarctica. In August Hurricane Harvey battered Houston and in September Hurricane Irma left a trail of destruction across the Caribbean. Climate change and its consequences for human existence are leading to interventions in water systems all over the world, with dikes being reinforced and coastal flood defences reassessed. For many countries the threat posed by the climate is new or, at most, only one of several challenges facing a government. In the Netherlands, by contrast, tackling the water has always determined the DNA of its inhabitants and public administration: the first *Hoogheemraadschap* (water authority) was founded in 1255; a national water authority with extensive powers, still called Rijkswaterstaat, has existed since 1798 and is currently part of the Ministry of Infrastructure and the Environment. This small country of just 41,000 km^2 also has water boards that organize regional water management. In 1950 there were still nearly 3,000 water boards but ongoing mergers have reduced the number to 21. These organizations, with their long history, make clear that water management is a public matter and in our collective interest – an interest that takes priority over the individual's interest because, without effective water management, the individual would literally be sunk. In 1857 the British historian Thomas Buckle wrote in the *History of Civilization in England* that 'the great division between European and non-European civilization turns on the fact that in Europe man is stronger than nature, and that elsewhere nature is stronger than man, the consequence of which is that in Europe alone has man subdued nature to his service.' The urgent collective need for a national system of water management has given the public administrative system in the Netherlands its comprehensive and decisive nature, with centralized economic and spatial development that has led to an unheard of level of prosperity. But the approach in which, as Buckle put it, nature serves man has proven to be untenable in recent decades. The results of climate change have made the engineers aware that new ways must be found to deal with natural systems and human interventions. And that is precisely what has happened in the Netherlands over the last 30 years. After 1945 a series of large-scale spatial planning interventions succeeded each other: the country's post-war reconstruction, then the overhaul of rural areas through land consolidation, next the Delta Works, followed by the expansion of an intricate distribution network consisting of motorways, navigation routes and airports and, finally, the implementation of a national ecological network. This string of interventions also reflects changing public sentiment, from an offensive approach to the landscape to an awareness of the need to balance the existing and the new. In recent decades the Dutch water sector has come to the conclusion that the water system's own patterns are the best teacher for protecting the country against flooding and ensuring its future. This change in thinking is remarkable for a country that has manipulated nature for centuries. Besides flood protection, this new approach to water provides other benefits such as habitat recovery, recreation and tourism, and the revitalization of cities and villages. Each change, including climate change, is an opportunity that is met with an open mind and the combined efforts of innumerable experts, providing a solution that is richer and more profound than a purely mechanical one.

This constant alertness led the OECD to award the Netherlands top marks in its important 2014 report on the country's water management, *Water Governance in the Netherlands: Fit for the Future?*. We seem to be in a permanent state of construction, as the next instalment in this long history is already in the pipeline: climate change is causing more heat stress and heavy downpours in urban areas, where it is difficult to contain

The Afsluitdijk that turned the Zuiderzee into an inland lake in 1932

The map shows the annual average of the heat in Amsterdam in relation to the environment. The difference between city and countryside remains below 3 degrees. On summer days the difference can be significantly larger

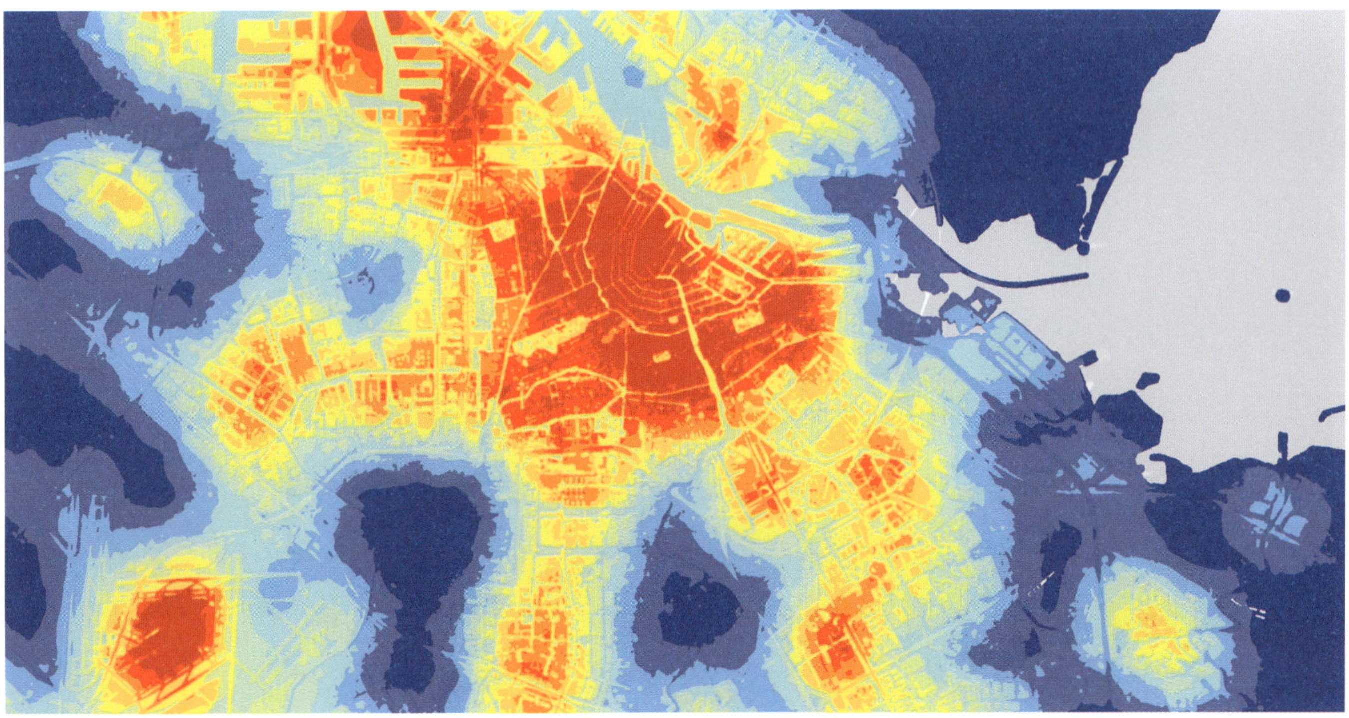

the water and temperatures may be as much as 7 degrees higher than in the surrounding rural area.

The centuries-old focus on flood protection has intensified further in recent years. Since 2010 provinces, water boards, municipalities and the Ministry of Infrastructure and the Environment have been working on the Delta Programme. An independent Delta Commissioner is in charge of the programme, which falls under the remit of the minister. The Deltafonds (Delta Fund) provides a secure financial basis for the programme, with an average annual budget of 1.2 billion euros for flood protection, fresh-water supplies and water quality. The role and operations of the Delta Programme, Delta Commissioner and Delta Fund were anchored in the 2012 Water Act by the 2011 Delta Act. The goal is to ensure that flood protection, the supply of fresh water and the spatial planning of the Netherlands are climate-proof and water-resilient in the long term, so that the country can continue to cope effectively with greater climate extremes. The 2014 Delta Decisions formulate clear interim goals for 2050 and contain specific research and implementation proposals for six issues: flood protection, fresh-water supplies, spatial adaptation, the Rhine-Meuse delta, the IJsselmeer, and coastal maintenance. This research and the solutions it produces has not been a central-government matter for some time now. Instead, it is carried out with other governmental levels, consultancies, construction companies and researchers. Preferential strategies describe the best approach and the governmental level at which it should be implemented.

The Interaction between Technology and Landscape Formation

In the long tradition of 'constructing' the Netherlands, man's use of technology has changed continuously. Looking back, we can distinguish different stages in the process. For centuries humans used technology to manipulate the water system without considering how this might affect the natural system. The goals of an intervention were flood defence and a manageable water level so that people could live in low-lying areas where they could farm and raise cattle. The gate culverts in the dams in Vlaardingen, dating from the third century BC, are a fine example. During high tide these gate culverts closed off the creeks that flowed inland from the sea, and when the tide was low they discharged excess water from the hinterland into the sea. In the north of the Netherlands people settled on artificial dwelling mounds known as *terpen* and *wierden* in pre-Christian times to escape the rising sea level and to farm the fertile sea clay. Technologies continued to be developed, refined and adapted for use in larger areas.

Reed Screener

Every year ten kilometres of reed screens are placed along the edge of the dunes to protect the coastline against erosion, a very labour-intensive undertaking. At the Hoogheemraadschap Hollands Noorderkwartier employees developed a machine in their spare time that can place an astonishing 600 metres of screen in 90 minutes. This means that six workers no longer spend a whole day on this task.

The polder windmills at Kinderdijk, a World Heritage site and skating paradise

But all that ingenuity also had a negative side-effect. In the west of the Netherlands, starting in the Middle Ages, digging drainage canals to create farmland caused subsidence of the peat soil, which oxidizes and becomes more compact when drained of water. Extracting salt from dug-up and burnt peat greatly exacerbated the problem, especially in Zeeland. Over the centuries the drained peat areas sank below sea level. When peat was extracted from the low-lying peat areas the rate of subsidence accelerated dramatically. Storms caused large pieces of peat to break off, creating vast unmanageable areas of water in the centre of Holland. The 'water wolf' was an urgent social problem and provided the impetus for a crucial innovation: windmills. Windmills made it possible to regulate the water level in the peat polders and to drain lakes. It is hardly surprising that the fifteenth-century Kinderdijk, with its nineteen famous polder windmills and drainage canals, has been listed as a UNESCO World Heritage site. The new land created in 1612 by the Beemster reclamation, another World Heritage site, used to be an vast peat lake. Draining the lake combined three interventions: a dangerous inland sea was conquered, fertile meadows were created and, in the midst of the well-fed cattle, country houses were constructed for Amsterdam investors. Located along the edge of the reclamation were mills that drained into the Ringvaart canal. The set-up of the pumping operation for the next reclamation, the Schermer, was even more innovative. Introducing an inner storage basin (northern and southern canals with smaller side canals) meant the polder could be divided into fourteen sections, each with its own polder level and mill that drained into the inner storage basin. Two large complexes of mills consisting of 38 mills then pumped the water out of the inner storage basin via temporary storage basins. This arrangement meant the mills worked optimally together even if one of them broke down. The system worked so well that the steam age passed unnoticed in the polder, and electricity was not introduced until 1925. The low-lying subsided peat area also played an important role in the construction of the Nieuwe Hollandse Waterlinie defence line. The Waterlinie is an 85-kilometre-long complex comprising locks, forts, dikes and inundation fields that were to protect the west of the Netherlands against the threat of war. Throughout the nineteenth century the country's best civil engineers worked on the Waterlinie. The vulnerable Zuiderzee dikes were also upgraded in the nineteenth century, made more urgent by a deluge in 1825. In 1833 the authoritative work on dikes, *Handleiding tot de kennis der dijksbouw en zeeweringskunde*, was written by the engineer Abraham Caland (1789–1869). He rejected the vulnerable high coastal dike, preferring a wide, gradually inclining dike that did not break the crashing waves but rather robbed the current of all its force because the water was forced upwards. His son, the engineer Pieter Caland (1826–1902), became world famous with his design for the Nieuwe Waterweg canal that was opened in 1872. Because the Meuse kept silting up, Caland proposed redirecting the winding Scheur river in a straight line to the sea, a distance of 4.3 kilometres. Caland's brilliant stroke was to let the river itself collaborate. He developed a civil engineering design in which the Meuse itself began to wear away its silted-up bed 30 kilometres west of Rotterdam. Locks near the sea were not needed and the discharge of the river ensured that the waterway remained deep. Ships first used this route to reach the sea in 1872. Steam ships, which became increasingly important at this time, made the Nieuwe Waterweg a good investment and transformed Rotterdam into a gateway to the world.

From about 1850 there was a growing realization that rivers could be kept at the required depth by removing sandbars, islands and floodplains to create uninterrupted navigation channels, by establishing a fixed width for river sections, and by constructing groynes to narrow the river bed in order to keep the navigation channel in place. This new realization shaped the approach taken by successive governments until the 1980s. It certainly helped, but steam dredgers had to be deployed to keep the navigation channels deep enough. The introduction of the steam pumping station turned the water engineer's world into an even greater miracle of precision and efficiency. The Woudagemaal in Friesland, which was put into operation in 1920, lowers the water level in the province's waters with the aid of a steam machine. This installation is also a UNESCO World Heritage site: it is the world's largest steam pumping station still in operation. The steam-driven centrifugal pumps move approximately 6 million m^3 of water a day. The Friesland water board has a permanent team in charge of the pumping station, which requires six hours to start and warm up.

Technical ingenuity and a pioneering spirit: the results were revolutionary and endowed the Netherlands even then with an outstanding reputation in the field of water management. The interventions in the waterways created the foundations on which the modern Netherlands could develop: shorter travel times and more efficient navigation. When an entire inland sea, the Zuiderzee, was successfully closed off with the enormous Afsluitdijk dam, the hydraulic engineer's authority became unquestionable. The water had been subjugated, the country could flourish: rivers were dammed up and dammed in, canalized, normalized, redirected and diked. Tricky bends were straightened, floodplains cleaned up. The architecture of the Hagestein dam and locks complex, which was completed in 1958, visibly expresses the transfer of forces, emphasizing pride in the functionality of water management.

The first careful steps towards a more adaptive approach to water management in the Netherlands were formulated by a biologist, not a hydraulic engineer In 1930 Wouter van Dieren (1902–1935) published his findings on the mechanisms of sand nourishment on Terschelling, one of the Wadden islands. Van Dieren studied historical maps, analysed plant communities, and researched the mineral content of the sand and the dispersal of salt by the sea wind. Central to van Dieren's work was coastal development as the driving force behind sand accretion and decomposition processes, as well as plant growth as an essential factor in dune formation. He emphasized the importance of incipient foredunes for preserving and renewing the dune area and argued that it 'is a misconception to constantly fixate them no matter how great the cost'. Van Dieren was a scientist who carried out interdisciplinary research into the natural and man-made aspects of the landscape. This approach enabled him to reach integral rather than sectoral conclusions. He died young but he pioneered a way of thinking that is now, nearly a hundred years later, the norm.

During the 1930s biology was joined by another discipline in the new debate about shaping the landscape and the way natural systems work. The landscape architect Jan Bijhouwer (1898–1974) intended to continue the tradition of developing the landscape. He had written his doctoral thesis on geo-botany and therefore had an unparalleled understanding of natural systems and their vegetation. In those years public debate raged about retaining valuable natural areas, but in his article 'Behouden of scheppen' (literally 'Preserve or Create') Bijhouwer pointed out that the whole of the west of the Netherlands is a created product, the result of human intervention and not of conservation. The landscape architect could design new landscapes by analysing landscape types and entering into a dialogue with urban planners, geographers, sociologists and civil engineers. Bijhouwer put his arguments into practice when he became a member of the Beplantingscommissie (plant committee) for the Noordoostpolder, which was reclaimed in 1942. A landscape was designed for the polder that suited the mechanized commercial farm of the twentieth century.

Undercurrent: The Delta Works

The overwhelming zest with which the post-war Netherlands was rebuilt and modernized did not extend to its coastal defences. The sea dikes were vulnerable and, despite repeated warnings, the government did not provide funding for their reinforcement. When the catastrophic North Sea Flood struck in February 1953, 1,800 people lost their lives. The government's swift response included the Delta Commission, Delta Act and Deltadienst. From 1956 the Deltadienst was in charge of the Delta Works, a system of thirteen dams and storm surge barriers that closed off broad, deep coastal inlets with their strong tides and currents.

Aided by a method of probability calculus that had already been developed in the 1930s and tidal calculations during storm tides, the Delta Committee no longer fixed dike heights determinatively (based on the highest mea-

Continuous monitoring

The Watermanagementcentrum Nederland (WMCN) is a crucial link in providing protection during flooding, storms and storm threats. Using the WMCN's expertise, Rijkswaterstaat works closely with the KNMI (Royal Netherlands Meteorological Institute), water authorities, provinces, Safety Regions and knowledge institutes. Combining knowledge and experience enables the WMCN to carry out its tasks more accurately and efficiently, no matter how critical the situation. The result: more and better reporting and predictions. Professionals have a single point of contact for matters related to current water information and water management. The Watermanagementcentrum in Lelystad issues warnings if the situation created by a storm or gales is so serious that measures must be taken. Staff measure the water levels along the coast, the major rivers and lakes continuously. The predictions are based on forecasts made by the KNMI. The situation is particularly tense during northwest storms with wind force 9 or higher: all the sea water is then blown towards the Netherlands, which can lead to very high water levels, especially if there is a spring tide. In these conditions, the KNMI sends a staff member to Lelystad for intensive consultations with the crisis team about the weather and water levels.

The Maaslantkering, completed in 1997, is situated in the Nieuwe Waterweg near Hoek van Holland

sured water level); instead, the still new science of probability theory was applied. In the Delta Plan, which appeared within a year, the maximum permissible chance of flooding in the densely populated Randstad was fixed at 1:10,000 per year. This norm was based on the fact that the Randstad would suffer the greatest economic damage in the event of a disaster. For the north and south-west of the Netherlands the Deltadienst recommended applying an 'economic reduction', fixing the norm at 1:4,000. Jan Tinbergen (1903–1994), a member of the Delta Committee and later winner of the Nobel Prize, introduced the cost–benefit analysis as an instrument. All waterways could be closed off except the Nieuwe Waterweg and the Westerschelde, which provided access to Rotterdam and Antwerp. The dikes here therefore had to meet the *deltahoogte* (minimum safe height).

The Delta Works as an Improved System

The Deltadienst was an outstanding organization for overseeing the Delta Works, with direct lines of communication with politicians and researchers. The organization's culture was characterized by horizontal knowledge structures and working with scenarios, which was completely out of step with the times. In the 25 years of its existence the Deltadienst constantly reinvented itself, encouraging new methods of calculation, new social claims such as ecology and environmental values, as well as new ways of working (scenarios). Closing off the coastal inlets, with their strong currents, would never have succeeded without the development of new techniques. The hydraulic engineers deliberately chose to build from small to large, from easy to difficult. That way the work itself would be a learning experience: new technical possibilities could be trialled and developed while the Delta Plan was being implemented. The series of dams and storm surge barriers is therefore an 'improved system': not only were innovations in civil and hydraulic engineering applied but the dams themselves also became testing grounds for landscape design, new natural environments, new beaches and leisure areas. With the aid of research carried out at the Waterloopkundig Laboratorium in Marknesse, and with Staatsbosbeheer as a partner for designing the new landscapes, a series of dams and barriers were built, each one an improvement on the last.

The Brouwersdam, completed in 1971, is still considered a masterpiece by Deltadienst employees. The whole area between the Brouwersdam and the Grevelingendam, in total 15,000 hectares of water and land, was designed as a nature and recreational area with national appeal. The high aspirations for the Brouwersdam resulted in a

A crowded Grevelingensluis, Zeeland

Waterloopbos

An experimental garden for big boys – that is how you could describe the Waterloopkundig Laboratorium (Water Cycle Laboratory) 'de Voorst' in Marknesse in the Noordoostpolder. In 1951, 36 permanent model locations were installed on the newly reclaimed land. In the course of 40 years, some 200 hydraulic models were constructed here. These were test arrangements for the large sea-arm ports and river improvements implemented in the Netherlands after the war such as the Delta Works. However, these test arrangements were also used for many hydraulic works abroad. In the late 1980s the site became obsolete due to digitization. In the years following, a unique cultural heritage site literally grew on the site, where remnants of technology now go hand in hand with nature values. In 2013 this Waterloopbos nature reserve was nominated as a 'top monument', an outstanding contribution to post-war reconstruction dating from the period between 1959 and 1965: 'de Voorst' was an indispensable link in the country's history of flood protection research. Natuurmonumenten and the Cultural Heritage Agency are working on a recovery plan for the Waterloopbos. To raise public awareness about water, RAAF studio and Atelier de Lyon have been commissioned to turn the former Deltagoot (Delta Flume), which was used to test the force of waves in the past, into a work of art. Ronald Rietveld of RAAF: 'By excavating the flume, a gigantic "delta work" to scale, 7 metres high and 250 metres long, will rise out of the surrounding water. Concrete slabs will be cut out of the thick walls and turned 90 degrees on their axis. The many different sizes and angles of the cut-out slabs will create a majestic space that opens up to the surrounding landscape and challenges visitors to embark on a journey of discovery.'
Address:
Voorsterweg 36, 8316 PT Marknesse

'landscape on the sea' that extends for nearly 6.5 kilometres. On the sea side of the Brouwersdam the slope was faced with asphalt, a clear reference to its damming function, with a sweeping beach at its foot that is really a continuation of the coast. The inner berm was greener and gentler, suited to the shore of a lake. For the first time, Rijkswaterstaat had dared to construct a road on the outer berm. The Brouwersdam is a publicly accessible landscape without equal.

The 'Planned Ecology' of Henk Saeijs, Biologist at the Deltadienst

Biologist Henk Saeijs (1935–2016) started working in 1971 as one of the first biologists for the Deltadienst's Afdeling Milieuonderzoek (environmental research department). He was initially received by the hydraulic engineers with scepticism but his ascent was rapid. In 1972 the department employed 36 people; by 1976 there were 83 employees. It was revolutionary, in hindsight, to designate the entire delta as a research area, from the Nieuwe Waterweg to the Westerschelde, including the North Sea up to 40 kilometres from the coast and the stream and river discharge area of west Brabant. Saeijs was given the veiled mission to convince the technocrats from within of the importance of learning about natural processes. He was the inventor of the system-oriented approach to water management, which centred on creating favourable pre-conditions in the Delta waters. Saeijs's discovery was that natural processes can be conditioned, something he tested in salt- and fresh-water conditions and every variant between the two: brackish (slightly salty) conditions, and tidal and mud-flat systems. He showed that when swamps and peat bogs were formed, they could encapsulate excessive nutrients so that polder water became clear. Creating new types of landscapes, formulated as a futuristic dream by Bijhouwer in 1934, became reality under Saeijs. He built *with* nature and when he retired in 1999 he was referred to as Rijkswaterstaat's first nature-engineer. The Grevelingenmeer was originally meant to be turned into a pure fresh-water lake that would provide water for agriculture. However, the disappearance of the tidal system and the slow desalination that would affect the existing ecosystem could not be reconciled with the general public's thinking at the time. It was therefore decided that the salt level in the Grevelingenmeer should be returned to what it had been before it was closed off by building a discharge sluice into the southern-most dam section. This created Europe's largest salt-water lake. These were years of experimentation. C. Janse, the engineer in charge of the research department of the south-west district of Rijkswaterstaat's Directie Watershuishouding, found this difficult to deal with:

The Brouwersdam, part of the Delta Works, a protective coastal defence and at the same time a recreation landscape

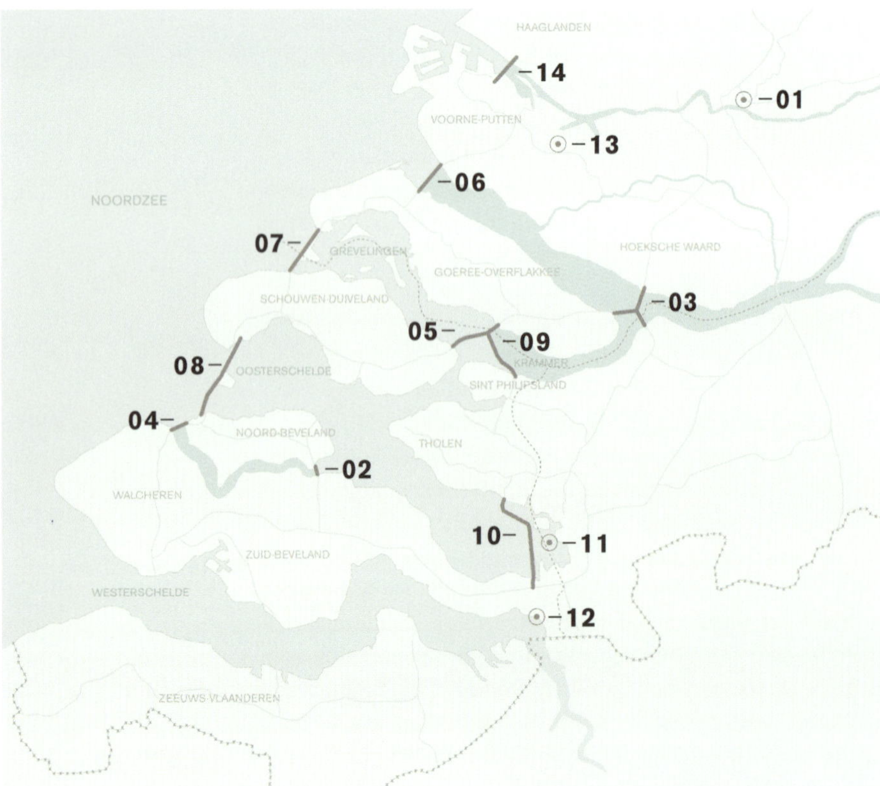

The Delta Works coastal flood defence system as it was implemented after the North Sea Flood in 1953

01 Algerakering
02 Zandkreekdam
03 Volkerakwerken
04 Veerse Gatdam
05 Grevelingendam
06 Haringvliet dam and floodgate
07 Brouwersdam
08 Oosterscheldekering
09 Philipsdam
10 Oesterdam
11 Markiezaatskade
12 Bathse Spuisluis and Spuikanaal
13 Hartelkering
14 Maeslantkering

'Using physics, we mastered the techniques to calculate water levels fairly well and that is why we use a mathematical model. But when other matters are concerned, such as salt penetration in the northern Delta basin, we are still unable to express that neatly in formulae ... You take the status quo, calculate it, wait ten years and then nature itself usually provides a setback', he said in 1984. Saeijs, by contrast, experimented with oysters and mussels that filtered away the unwanted algae that resulted from excessive fresh water, and with fields of sea grass that suppressed the growth of algae. He introduced concepts such as 'eco-technique' and 'planned ecology'. As a biologist in the midst of hydraulic engineers, he laid the basis for a new orientation towards integrated water management and, together with Minister of Transport, Public Works and Water Management Neelie Kroes, he made this into a politically anchored concept. Saeijs wrote in 1987: 'Adapting our surroundings to meet man's needs is not wrong in itself. What is wrong is the anthropocentric and opportunistic manner in which this is done. It is not change that needs to be resisted – change is, after all, the basis of life. What matters is making change happen in the right way.' The approach advocated by Saeijs was viewed with suspicion at the time but has now been embraced as 'building with nature'. In his autobiography Saeijs wrote about the very sensitive political debate about whether to leave the Oosterschelde open or not. By the end of 1973, 5 kilometres of the 8-kilometre-long sea arm had already been dammed. With ongoing demonstrations by commercial fishermen, shellfish farmers, marine yachtsmen and conservation organizations, the PPR party, which had joined the coalition government led by Johannes Den Uyl, threatened to abandon the government. This had to be prevented at all costs, so construction work was temporarily halted. In 1976 the government decided to build a half-open storm surge barrier that could be closed when necessary. It was completed in 1986. On an unstable bed in 40-metre-deep, violently swirling water a permeable barrier was built that remains unequalled anywhere in the world. The government's decision marked the breakthrough for Saeijs's thinking.

Creative Synthesis

The construction of the Delta Works was not only an integral approach to hydraulic engineering and ecology. It also heralded a new future for the south-west of the Netherlands. The Minister of Reconstruction and Public Housing Herman Witte (1909–1973), who was a qualified civil engineer, promoted the integration of agriculture; industrialization (the younger generation had a reason to stay); recreation (the fresh-water basins created economic opportunities); infrastructure (a recreational and port route); and social development (Zeeland was economically deprived). Witte, who was minister from 1952 until 1959, was convinced that the expansion of the post-war Netherlands could only succeed if material reconstruction was combined with a programme for cultural development. The Delta Works express the Dutch 'can-do mentality'. And this makes the works so much more than only protection against the dangerous sea. They are a crucial link in the international reputation of the Netherlands as a nation of civil and hydraulic engineers. In 2013 the International Federation of Consulting Engineers proclaimed the Delta Works the most prestigious civil engineering project in the world. The Delta Works unite the 'hard' world of civil engineering and the 'soft' world of recreational landscapes in a systematic construction of unparalleled harmony. Jasper Vink, the director of the Rijksdienst voor het Nationale Plan that coordinated post-war reconstruction at a national level, called this 'creative synthesis'. Under his leadership this regional approach was adopted for all regions in the Netherlands. The Delta Works are not designatedUNESCO World Heritage sites but they indisputably deserve to be, combining as they do technology and societal enrichment.

General in Times of War – Unrest in the River Areas

When the Brouwersdam was completed in 1971, social unrest was growing in the river areas. The norm for the crest height of river dikes had always been the highest water level measured plus 1 metre. River dikes were therefore raised after every flood but they were not widened, which meant they were quite steep. After the Great North Sea Flood in 1953 norms were established for river dikes, just as they had been for coastal defences: a one-in-3,000-year chance of flooding was considered acceptable. A large number of dikes did not conform to this norm and were therefore raised, reinforced and their bases widened by up to twice their original width. 'The *dijkgraaf* (dike reeve) is behaving like a general in times of war' wrote one Gorkum biologist. All the dwellings on the outer slope of the dike in the village of Brakel were demolished in 1974 for the sake of dike reinforcement. The villages of Dalem and Puttershoek were also virtually demolished. At the beginning of the twentieth century the nature studies teacher Jac. P. Thijsse had made the river areas famous among the general public through the Verkade albums: when people bought biscuits made by the Verkade factory they received several pictures painted by Jan Voerman jr. (1890–1976) that could be glued into these albums. Now it was those very landscapes that were threatened with extinction. Rijkswaterstaat's rigid approach and unassailable authority led to demonstrations. In 1975, following protests in the Dutch parliament by socialists and liberals, Minister of Transport, Public Works and Water Management Theodorus Westerterp appointed the Becht committee. Its remit was to investigate alternatives to dike reinforcement. The committee's recommendations included a revision of the norm for dike improvement: the chance of flooding for river dikes should be 1:1,250 years instead of 1:3,000 years. In transition areas, where sea and river influences come together, the norm was changed from 1:2,000 to 1:3,000. Greater attention was paid to landscape and cultural-historical values when balancing the various interests. The Becht committee argued for 'sophisticated designs' that suited the landscape and took account of historic dike buildings. The government and parliament accepted the committee's recommendations. A more equal treatment of natural values was also included in other policy areas. The 1975 Relatienota memorandum aimed to bridge the gap between agriculture and nature. This integration process was never successfully completed, at least, nowhere near as successfully as it was – albeit long after the Becht committee's report – for water.

'With that irritating river you never know what is going to happen next.'

From 1975 the approach to dike reinforcement was therefore supposed to be more nuanced, but nothing could have been further from the truth. The government had to cut spending, and the damage to its reputation following budget overruns for the Oosterscheldekering (15 per cent after correction for inflation), its prioritization of sea dikes, and the protection of the port of Rotterdam led it to scrap the agenda item 'dikes in the sparsely populated river areas'. In 1983 Minister of Transport, Public Works and Water Management Neelie Kroes suddenly presented new details concerning dike reinforcement: in both the upper and lower river areas the dikes were to be raised by up to 1 metre. Once more, there was widespread public discontent. It was thought that Rijkswaterstaat wanted to use the new heights to reverse the findings of the Becht committee – for example, the 1:1,250 years norm was raised to 1:4,000 for the Alblasserwaard. To top it all off, it was said that Rijkswaterstaat itself had caused the need for dike raising. The *NRC Handelsblad* newspaper wrote in 1984: 'Over the last fifteen years the Haringvliet and the Volkerak have been closed off with locks, the canalization of the Rhine has been completed, the Hartelkanaal opened and the Dordtse Kil widened. All this has reduced the major rivers' flow rate. This has, in turn, resulted in deposits of unforeseen quantities of sediment on the riverbeds.' In short, because the effects of the ebb and flow of tides no longer penetrated deep inland, the water levels were higher, which meant that the dikes had to be made higher. The effect was not noticeable in the lower rivers, below Werkendam, but in the Waal river it was unmistakable. A.J. Struijk, the engineer in charge of Rijkswaterstaat's Afdeling Waterbeheer van de Directie Benedenrivieren, which was responsible for managing the lower rivers, complained in the same article that 'With that irritating river you never know what is going to happen next.' The 'irritating river', the utterance of an engineer who could think only in terms of modelling reality on his own terms. The river, nature, they were the enemy that simply would not surrender. The reflexive response was dike reinforcement, estimated to cost 100 million guilders, but there was no discussion of what was causing the higher water levels, the shallowness and the ridges in the riverbeds. Nor was there any mention of a fatal side-effect: the polluted sludge had turned the Netherlands into north-western Europe's chemical dump – in some places the layer of sludge was several metres thick. It was not only a threat to public health; the riverbeds themselves became depleted, making it impossible for them to fulfil their vital functions.

Towards a Multi-Focused Water Management Policy

Thanks in part to the earlier contacts between the biologist Henk Saeijs and the minister, Neelie Kroes, parliament became convinced of the need for a different type of water policy. By then, the destruction caused by Rijkswaterstaat's inflexible and technocratic solutions had become unacceptable to the general public. Water policy became multi-focused; the 1984 Nota Waterhuishouding, a memorandum on water management, referred specifically to the need to balance social considerations. In 1985, 70 houses in Sliedrecht were preserved using hydraulic techniques, steeper slopes and dam walls.

At the Rijksplanologische Dienst, the government planning authority, the idea of 'creative synthesis' for larger areas, which had made the Delta Works so successful from a socio-economic perspective, was taken further by a new generation. It was adopted as the main aim of the Eo Wijers foundation, named after the urban planner and director of the Rijksplanologische Dienst who died in 1982. In 1985 the topic of the foundation's first design competition was 'Nederland Rivierenland' ('the Netherlands, land of rivers'). The foundation's directors thought it essential to tackle this intractable question, which had been dragging on for ten years, from multiple perspectives. One of the winning designs was the remarkable Plan Ooievaar: de toekomst van het rivierengebied ('Plan Stork: the future of the river areas'). It was drawn up by six young bureaucrats working at different ministeries – five at the Ministry of Agriculture and Fisheries (four landscape architects at Staatsbosbeheer and a biologist at the ministry), and one at Rijkswaterstaat as a project manager. Plan Ooievaar was a fresh look at the integration of nature, excavation, agriculture and river management. Its core themes included supporting agriculture on the heavy clay soils on the landside of the dikes and rolling back agriculture in the floodplains so that the summer dikes could be breached and the dynamics of the rivers harnessed to reinstate nature (riparian woodlands). With its simultaneous attention to apparently contradictory interests, the plan was a breath of fresh air for the interest groups hiding in the trenches. It was taken seriously by Rijkswaterstaat, partly because Neelie Kroes knew and respected one of the designers, Dick de Bruin, who worked at Rijkswaterstaat. She had just visited India with de Bruin to sign contracts for the use of Dutch dredgers in the Ganges delta, which was plagued by flooding. This project was managed by H. Ferguson, the head of engineering and director of the Deltadienst. The export value of Dutch water expertise has a long history.

Plan Ooievaar took a systems approach: the designers dreamed of a new balance for the river landscape: lowering floodplains would create the right conditions for riparian

Plan Ooievaar for the river areas, 1987

Coupure in the Diefdijk

Above the A2 motorway, just west of Culemborg, is a very special viaduct. Cars and bicycles use the viaduct's surface to cross over the motorway but, concealed under this surface, is a secret weapon. Three enormous concrete beams are stored under the viaduct behind its streamlined profile: they can be used to close off the space under the viaduct if there is a threat of flooding. The A2 motorway is, in fact, a coupure (cutting) in the centuries-old Diefdijk, which protects the low-lying Alblasserwaard and Vijfheerenlanden from flooding. If the coupure in the motorway is closed off, the water can flow away via spillways in the river dike at Gorinchem as soon as the water level in the Merwede river has dropped sufficiently (see p. 80).

When the dikes break

The television series Als de dijken breken ('When the dikes break') was broadcast in 2017 on Saturdays evenings. It was watched by 1.7 million viewers and many people responded on social media. National awareness of the threat posed by the water, in slumber-mode since the completion of the Delta Works, needed a shake-up. On the overstroomik.nl site the Dutch can type in their postcode to see how high the water level in their neighbourhood would be if there were a flood. Citizens are given advice about whether to leave or stay put in the event of high water. 'During a flood you can no longer see where you are driving or walking. If you get stuck on the motorway and are caught out by the rising water, you will become a victim. Sometimes it is better to stay put. If you camp in the attic, make sure you have enough drinking water, a transistor radio with new batteries, torches, candles, matches, blankets, dry towels and your ID.'

woodlands, flowing channels, dynamic river processes and meadows. Flora and fauna that were new or that had long ago disappeared, such as the black stork after which the plan was named, would return to the river areas. In 2015 Dirk Sijmons, one of the designers, described the development of Staatsbosbeheer's *casco* (frame) concept at the time: how could different sorts of land use – floodplain nature and agriculture; recreational areas and nature reserves – each with its own dynamic and annual cycle, be combined in a single area? The solution turned out to be a stable framework containing a strict separation between floodplains and agricultural areas, with room for flexible use within each separate area. The Ooievaar team was inspired by Saeijs's planned ecology and Wijers's integrated socio-economic approach as well as by the work of the Scottish-American professor of landscape architecture Ian McHarg. McHarg's work was propagated by Meto Vroom, professor of landscape architecture at Wageningen University. McHarg's *Design with Nature*, his pioneering 1969 study in ecological planning, was developed further by the department of landscape architecture at Wageningen University.

A series of political decisions supported the approach adopted in Plan Ooievaar. The semi-open Oosterscheldekering, which was nearing completion, symbolized a change in culture from a 'hard' water infrastructure to greater appreciation and space for the natural environment. A debate about the Oostvaardersplassen in the new Flevopolders was in full swing: a primaeval biotope – a fresh-water swamp with a river forest – had grown spontaneously on 6,000 hectares of land that had been reserved for industry but had remained vacant. This biotope was eventually considered so valuable that the Oostvaardersplassen became a protected site under the 1986 Natuurbeschermingswet (Nature Conservancy Act). This proved that nature was capable of producing new values based on conditions created by humankind. The Ministry of Agriculture was far from enthusiastic, however. Despite the 1975 Relatienota, a memorandum aimed at balancing natural, landscape and agricultural values, and despite – or, in fact because of – the fact that Plan Ooievaar was submitted by five fellow bureaucrats, the ministry's heads followed the logic of agricultural production and took no account of natural values.

The ideas underpinning Plan Ooievaar nevertheless got off to a flying start and were given an extra impulse by the 1986 Ontwerpwet op de waterhuishouding, a draft law on water management. The draft law was mainly concerned with the development, functioning and protection of the water systems but it also stated that these aims were inextricably linked with the required protection of these systems. The draft law evidenced an acceptance that integral water policy is a process of continuous transformation, with room for not only flood protection but also for new uses and ecological functions. In November 1986 Neelie Kroes ordered several summer dikes to be breached in Gelderland as an experiment. In 1987, looking back on his work at the Delta works, Saeijs called 'the creation of favourable conditions' the Art of the Nation. 'In recent decades, spectacular developments have taken place that are leading to a new approach to water systems, a type of planned ecology; a modern way of planning and managing cultural landscapes.' Saeijs saw that this new path was being followed throughout the Netherlands and that it required a united effort from central government, provinces, water boards and municipalities: 'All the governments that are involved must participate in a harmonized management, not only the [central] government.' They proved to be prophetic words.

Guts and Vision

A change in culture does not happen overnight but at the end of the 1980s a new dynamism in landscape architecture, water infrastructure and ecology was unmistakeable. Design studies such as the 1988 *Een scherpe grens. Ontwerpstudie naar de ruimtelijke kwaliteit van verzwaarde rivierdijken*, one of whose authors was the landscape designer for the Brouwersdam, clearly showed the potential of a different approach to dike reinforcement. Minister Neelie Kroes was personally involved in realizing the ideas contained in Plan Ooievaar. Kees Poot, a policy official at the Ministry of Transport, Public Works and Water Management at the time, recalls being instructed by Kroes to organize an excursion for members of parliament to the Millingerwaard. This is where the Rhine enters the Netherlands and separates into the Waal, Lower Rhine and IJssel rivers. All plants and animals that settle here, at the top of the Dutch delta, disperse with the current to the north and west. The minister was delighted with the results of the new approach and in 1989 even threatened to take severe measures to promote it: 'We will not hesitate to expropriate floodplains – as an extreme measure – within the framework of Plan Ooievaar. Otherwise we will never achieve our goal. The plan requires perseverance. It requires guts and vision to restore the ecological system that used to exist in Netherlands' river landscape to its former glory. It is not important who does it, what matters is that it gets done.' During an earlier boat excursion on the Waal she was told about the opportunities created by nature development and she saw how the irreplaceable Millingerduin was destroyed by excavators. 'Who owns that land and those excavation works?' inquired the minister. 'They are the government's, your own ministry's', came the reply. Upon hearing this, Kroes made sure that the World Wildlife Fund could start the Levende Rivieren ('living rivers') plan here in 1992, turning Plan Ooievaar's more natural approach to the river into government policy. This prompted the Plan Ooievaar designers to leave their government jobs and set up their own consultancies so they could continue working out their ideas in concrete projects. The Levende Rivieren plan investigated life in the riverbed. It disarmed even the greatest sceptics because it turned out that the 'soft' environmental sector actually possessed earning power – the excavations supplied large quantities of clay to the brick industry. Lowered floodplains can drain more water when water levels are high, thus reducing the risk of dikes collapsing. Carefully scraping 2 to 3 metres of clay off the sandy floor of the floodplains would return a landscape with secondary channels and sandy levees where nature itself would ensure a rapid revival as fertile soil was replenished and the river provided a continuous supply of seeds and organisms. The familiar grassy floodplains and wide river flowing past banks with groynes where it was deep enough for navigation would acquire the primaeval character of the lowland rivers, with their many meandering, shallow channels coursing along gravel banks, river dunes, sandy shores, riparian woodlands, grasslands and swamps. Changes were also afoot for coastal policy. A 1992 information video made by the Technische Adviescommissie voor de Waterkeringen talks about 'giving greater consideration to living and social conditions, nature and cultural history'. It also refers to the principle of 'dynamic maintenance', whereby sand nourishment is used for coastal reinforcement. According to the film, the coast meets the Delta norms, but this is far from being the case in the river areas.

'Things will never be right again along this dike'

Despite the early success of the new approach in the floodplains, Rijkswaterstaat's usual river-reinforcement practices changed at a snail's pace. In 1993 the *NRC Handelsblad* newspaper gave extensive coverage to the destruction of the village of Hellouw and the surrounding cultural landscape, which came straight out of the Verkade albums. When the dike was reinforced in the late 1980s the orchards and many old buildings were destroyed, despite proposals for the village to become a protected historic site. 'The steep, winding dike has been changed into a steep, unwieldy, concrete-and-basalt monster.' The river painter and activist Willem den Ouden (1928) lamented that 'Things will never be right again along this dike. Where Attila has been, nothing will grow anymore. The soul has been cut out of this landscape.' A degree of cynicism was not uncommon at Rijkswaterstaat: it often tackled dike reinforcements in ways that meant it could avoid submitting the mandatory MER environmental impact report. As a result, no uniform method was adopted and each section of a dike was dealt with differently, giving the landscape a disjointed aspect. Hellouw provided the impetus for yet another advisory committee, this one chaired by Cornelis Boertien, a former Queen's Commissioner of the Province of Zeeland. The committee examined the basis for reinforcing the river dikes, just as the Becht committee had done in 1975. Now, in 1993, support among the general public did exist.Minister of Transport, Public

Works and Water Management Johanna Maij-Weggen reacted positively to the committee's findings. It concluded that the reinforcements that had been implemented had badly affected natural, landscape and cultural-historical values. Bespoke solutions were now promised. In the future an environmental impact report would be mandatory for all sections of dikes, regardless of their length. Height and width norms for dikes would be more flexible, reverting to the Becht committee's 1:1, 250, thus protecting natural, landscape and cultural-historical values. In its advice, which was partly written by the landscape architect Professor Meta Vroom, who was referred to above, the brand new ideas about lowering dikes by scraping some of the soil off the floodplains were linked to the development of the natural environment and 'sophisticated' designs. Although the 'Levende rivieren' plan would then cost an extra 185 million guilders, the damage to the natural, landscape and cultural-historical values would be reduced by 55 per cent. In the future, landscape architects would be involved in designing dikes. The committee concluded that 'Until now, there has been insufficient legal protection, democratic involvement and openness.'

Reality Hits

In December 1993 severe flooding along the Meuse in Limburg and extremely high water levels in the Biesbosch region caught people unawares. The A2 motorway near 's-Hertogenbosch was flooded because the Dieze river could not discharge into the Meuse – the water level in the Meuse was higher than the Dieze. Motorway exits in the area had to be closed because of 'high-water tourism'. It was considered an incident. But then came January 1995: the water levels of not only the Meuse but also the Rhine were extremely high and the dikes were threatening to break. A crisis centre was set up in Arnhem, mayors wearing wellington boots appeared on television, and the Queen's Commissioner for Gelderland Jan Terlouw really did have to behave like a general in time of war, flying over the stricken area in a helicopter. It was he who took the difficult decision to evacuate 250,000 people. In 2017 Terlouw said, 'I entered the [evacuated] area with the Queen. It was bizarre, an empty village like that, completely intact, with the front-door lights still on. The evacuation was very disruptive for the inhabitants but it went perfectly. The Dutch are good at improvising. We deliberately let the Wilpsedijk be flooded. I stood on the dike with the dike reeve and we decided: this is a water storage area, we don't need to resist the water here at all costs.' Politicians were quick to blame environmentalists for the floods – they had put the dike landscape above inhabitants' safety. Although they were silenced by Hans Dijkstal and Annemarie Jorritsma – both members of his own party and government ministers – initial public support for those who had opposed the rigorous dike reinforcements turned into rage. The painter Willem den Ouden was threatened and even shot at. (Much can change in 25 years – the recent redevelopment of the IJssel river near Zwolle was based on a painting by Jan Voerman (1857–1941).) The effect of the 1995 near-disaster was that, just as with the Great North Sea Flood in 1953, more manpower and resources were deployed to tackle weak spots. And just as in 1953, government bodies reinvented themselves, driven not by a single, centralized Deltadienst but by Staatsbosbeheer and Rijkswaterstaat, which built on the systems approach and brought together hydraulic engineering, landscape architecture and nature development in the new regional approach symbolized by Plan Ooievaar. The new measures were of course laid down in a series of policy plans, acts, memoranda and pilot studies. Certainty is essential when it concerns security measures whose potential consequences for the public are so great. September 1995 marked the beginning of the Delta Plan Grote Rivieren (Delta Plan for the Major Rivers), which was anchored in the 1995 Delta Act. The first projects also began in 1995, including the Gamerensewaard near Zaltbommel, where large new natural channels were created. In 1996 the Wet op de waterkering (Flood Defences Act) was passed, followed by the Beleidslijn Ruimte voor de Rivier, a policy paper on 'Room for the River'. In 1998 the Vierde nota waterhuishouding, a memorandum on water management, was issued, expressing a preference for spatial rather than technical (dike-raising) measures. It was based on the 'Yes, but if . . . ' principle: if a river will have less room due to an intervention, then secondary channels must be dug elsewhere as compensation. In 2000 it was time for the next advisory committee, this one chaired by Frans Tielrooij, who expressed his thoughts on the type of water management that was required for the twenty-first century. His central message was that the water needed more room. In the foreword to the committee's report, he wrote, 'It is indisputable that the Netherlands, the drain of Europe, requires a different approach. There is still too much talk of technical management, whereas it is high time for a different water policy. A water policy that is broadly oriented, focusing on content, public support and direction, and that contains clearly formulated responsibilities.'

High water in the river areas. 250,000 people were evacuated as a precautionary measure, 1995

It is fascinating to see how much energy and brainpower were invested in developing preliminary studies, research projects, models and simulations in the search for new technical norms after the 1995 floods. For example, Rijkswaterstaat commissioned RIZA, the government body charged with managing the country's fresh-water supply, to develop a computer programme for the river landscape of the Rhine tributaries. It was called the *blokkendoos* ('box of building blocks') and was used to investigate which normative water levels would occur along the Meuse and different sections (upper, lower and so on) of the Merwede when 15,000, 16,000 and 18,000 m^3 per second were being discharged. It was then possible to see the effect of proposed interventions on the water level at Gorinchem or Werkendam. In 2000 it was decided that the Noordwaard and Overdiepse polders would be turned into overflow areas (see projects on pp. 132–135 and 140–145), This was followed by the 2001 Planologische Kernbeslissing 'Ruimte voor de Rivier' (Key Planning Decision 'Room for the

Dirty Meuse water

Inge Lardinois, head of the water policy programming department at the Ministry of Infrastructure and the Environment: 'My father is from Itteren, a village along the Meuse, which was flooded several times in the past century. My aunt told me that every time the sewage water flowed in first, followed by the dirty Meuse water, causing much damage and inconvenience. In the summer of 1980 the village of Itteren was inundated too: the Meuse had overflowed its banks again. I visited my aunt's house with my father and brother; my grandparents' farm, which was very near the Meuse, was inaccessible but it had stayed dry because it was one metre higher.'

Sliding down the Hollandse IJsseldijk in Capelle aan den IJssel

Deltagoot

The wave-breaking effect of the Afsluitdijk, or the strength of the asphalt used in the Waddenzee dikes: these are the types of flood defence tests carried out in the Deltagoot (Delta Flume) at the Deltares research institute. This flume, 300 metres long, 5 metres wide and nearly 10 metres deep, simulates extreme weather conditions to test the effects of waves on dikes and dunes. Paul van Steeg of Deltares says of the Waddenzee dikes that 'Asphalt is a strong dike revetment material but we do not know yet how strong asphalt is when it is 30 to 50 years old. This test in the Delta Flume will enable us to find that out. Using these tests, we will be able to see whether the dike asphalt can last longer than currently calculated based on our calculation models. That could save a lot of money.' Ten strips of asphalt from the Lauwersmeerdijk were removed from the dike in the summer of 2017 by the Wetterskip Fryslân water authority and placed on the simulated dike in the Delta Flume. High waves were used to simulate and monitor the subsidence process of the asphalt. This will provide new insights and knowledge for dike improvement in the Netherlands. The experiment is an opportunity to evaluate and improve the assessment criteria applied to asphalt revetments. This might make future dike reinforcements cheaper. However, even with the wealth of computer models currently available, physical trials using real water and real waves remain irreplaceable.

River'). The crux of this Key Planning Decision, with absolute powers of implementation, was its dual goal of flood protection *and* guaranteeing spatial quality. (According to the former programme director Ingwer de Boer, 'Policy and implementation belong together. But separating Rijkswaterstaat's implementation division and the water policy division in 2003 meant this was no longer self-evident. So that is what I pushed for with the two directors-general. After all, the three of us were working for the same minister.') From 2005 all projects were tested by the Kwaliteitsteam Ruimte voor de Rivier (Q-team Room for the River), which controls spatial quality. It goes without saying that in the world of water infrastructure, spatial quality had to be measurable, and that was no easy task. The first chairman of the quality-control team, Dirk Sijmons, says: 'The Rekenkamer (Court of Audit) posed the question: how we can make spatial quality measurable? Well, we used before, between and after photographs, which were presented in folders.'

The river improvement programme Ruimte voor de Rivier (Room for the River) comprises 34 projects and has now been more or less completed. It cost 2.4 billion euros, and the changed aspect of the river areas is obvious to any observer. The picture sketched by the World Wildlife Fund in the Levende Rivieren plan has not been fully realized. The engineer and top Rijkswaterstaat adviser Tjalle de Haan: 'The river landscape after Room for the River fortunately achieved a balance between the Living Rivers concept and that of river normalization that was completed in about 1900. This normalization consisted of removing all sorts of islands in the summer bed and obstacles such as woods from the winter bed so that when the thaw set in, ice would no longer get caught and create ice dams that dammed up the water. These had been the main cause of flooding for centuries. Besides this, the discharge between groynes during low water would be concentrated to guarantee sufficient depth for navigation.'

The fruits and innovation of Dutch water works, from the Afsluitdijk via the Deltadienst and Plan Ooievaar to new policy on rivers illustrates perfectly the proposal put forward by the Italian-American professor of innovation economics Mariana Mazzucato in her 2013 book *The Entrepreneurial State: Debunking Public vs. Private Sector Myths*: government plays a much greater role in innovation than is often admitted. Her message: 'The private sector only finds the courage to invest after an entrepreneurial state has made the high-risk investments.' Without government-facilitated research the iPhone would not exist, as she explains in her books and lectures.

Rijkswaterstaat Changes

The impressive chain of risk management, responsibilities and strategies that we now use in the water sector has a long history, as does the approach integrating technology, natural development and economic earning power. The Department of Hydraulic Engineering at Delft University of Technology, with its close ties to Rijkswaterstaat, has played a key role in developments. Han Vrijling, then professor of probabilistic design and hydraulic structures at the university, began his career working on the Oosterscheldekering. He is considered the founder of probabilistic design, which is now accepted as the norm. This approach considers not only the chance of high water but also the chance that all sorts of circumstances might combine to put unsustainable pressure on a dike. This observation led Vrijling and his team to develop a probabilistic estimation and planning method that could also be used for risk analysis for road traffic, spatial design and air pollution. (The Bijlmer air disaster in 1992, when an El Al Boeing crashed into a tower block in Amsterdam, and the fire in the Channel Tunnel in 1996 reinforced this development.) Besides technical measures, the method also dealt with contingency plans and emergency relief in order to limit the effects of possible failures: the risk-based approach. This type of research signalled a revolutionary extension of Rijkswaterstaat's work, implemented in close cooperation with Delft University of Technology in the Technische Adviescommissie Waterkeringen, a technical advisory committee. The application of computer techniques, systems engineering, model studies and policy analysis techniques created a new way of thinking about risks.

A second turning point was signalled in 1988 by Han Vrijling's colleague, Professor Hennes de Ridder, a former designer who had also worked on the Oosterscheldekering. The role of Rijkswaterstaat had to change: it should be less of an actor and become more of a director. 'One of the most important pre-conditions for collaboration is that business must not only enjoy the benefits of its surroundings but it must also bear the burdens of its surroundings. That is new. The way risks are going to be distributed is also new. The risks of implementing and operating the bare bones of essential projects will be borne by Rijkswaterstaat. The risks of implementing and operating the non-essential amenities will be borne by companies that will enter contracts to do so.' Here was a new sphere of action: the government was to balance and define public versus private interests, and technical solutions versus experiential values.

Upstream: The Integral Approach to Projects and Responsibilities

'From the start of my time in office I have resisted the segregated, sectoral approach to projects', said Tineke Huizinga, who was State Secretary for Transport, Public Works and Water Management from 2007 until 2010 in the coalition government led by Jan Peter Balkenende. 'I wanted a programme for flood protection that would remain outside the political ball game', she told the *Groene Amsterdammer* magazine in 2015. 'A programme that would organize our flood protection for the long term and that was not subject to an unquestioning acceptance of the status quo.' The water sector has once again reinvented itself. The Delta Committee's recommendations in 2008, the start of the Delta Programme and the appointment of an independent Delta Commissioner in 2010, as well as the lessons learnt from Room for the River enabled the Ministry of Infrastructure and the Environment and its implementing body, Rijkswaterstaat, to shape their new role: from actors to directors. Revising the flood norms was at the core of Minister of Infrastructure and the Environment Melanie Schultz van Haegen's time in office. Matthijs Kok, Professor of Flood Risk at Delft University of Technology and involved in setting up the new norms, told the *Groene Amsterdammer* in 2015: 'We had known for some time that we had to make a transition from a uniform norm within a dike ring to a differentiated policy. Applying the same safety standards to a dike that protects two meadows and a few cows and a dike that protects 500,000 people is not logical.'

It sounds harsh, but in order to achieve the right balance for each location a cost–benefit analysis is used whose results are translated into money. A human life is a loss amounting to 5.6 million euros, an evacuee costs 10,000 euros. And this is just the beginning. From homes, offices, hospitals and infrastructure to the economic damage if people do not work for days, systems fail, electricity is cut, the list is endless. 'It is an expertise', says Kok, who has specialized in these kinds of risk analyses. 'The facts and figures are useful but they do not determine the acceptable risk; the norms are the result of balancing considerations.' The new norms were fixed as of 1 January 2017 in the Water Act, crowning the minister's efforts. Everyone in the Netherlands is now covered by the *basisbeschermingsniveau* ('basic level of protection'): the chance that someone might die as a result of a flood must not be greater than one in a hundred thousand years. Areas that would suffer widespread social disruption, with a high death toll and extensive economic damage, are given extra protection.

Two-Way Traffic

Richard Jorissen, director of the Hoogwaterbeschermingsprogramma (Flood Protection Programme), began his career at the Deltadienst in 1985. Describing the recent approach taken by Rijkswaterstaat, he explains that any matters that used to be taken for granted have changed in recent years. Those who formulate water policy are no longer involved in its implementation. The usual government funding of projects has been replaced with joint financing with the water boards. Spatial quality, which has a legal basis in the Key Planning Decision Room for the River, is not always enforceable in later programmes but it must still be addressed. These new arrangements have ensured that the different parties in the water sector must actively seek out each other as well as the 'users' of their end-

Water light

Since February 2015 Studio Roosegaarde's 'Waterlicht' installation has been touring the Netherlands. Dijkgraaf (dike reeve) Hein Pieper of the Waterschap Rijn en IJssel water authority had the work premiered in the floodplains near Westervoort to show what the Netherlands would look like without dikes. Wavy lines of blue LED light in the night sky allow visitors to experience how high the water would be without human intervention. The lines of light shine through each other around the chosen location while they slowly move up and down. 'Waterlicht' is a virtual flood and aims to increase awareness of the constant danger of water. The power of this enchanting, poetic experience is putting the Netherlands as a work of art on the map.

product, the citizens of the Netherlands.
New life has recently been blown into old relationships. Rijkswaterstaat and Staatsbosbeheer had collaborated since 1916 on major projects such as the Rijkswegenplan (a road network plan), land consolidation and the Delta Works. The two bodies recently reached a cooperation agreement to supplement and reinforce each other's work in planning programmes, management and conservation, sustainable energy, landscape and cultural heritage, and expertise and innovation. Expertise and innovation were given particular emphasis because the separation of policy and implementation risks fragmenting expertise. What expertise do the organizations have and where are the gaps? How can expertise be shared, and how can systems expertise be linked to integrated management? Saeijs had already warned that 'Research must form an integral part of project management, from start to finish. There must be room for technical adjustments, even after the actual project has been completed. The project is, after all, not finished when the works have been completed.' But it is not only about the expertise and experience gained from the projects themselves but also about critically evaluating established policy goals as formulated by the Ministry of Infrastructure and the Environment: what works, and what doesn't?

Beyond the Dikes

The Dutch water sector is flourishing, that much is clear. Annually recalibrating the Delta Programme ensures a clear overview of planning and budgets. What is currently of interest is the sector's greatly increasing scope due to two factors. First, the concept of multi-layer flood risk management: besides dike reinforcement, a broader approach to flood risks is being taken. Layer 1 deals with the flood-control barriers, layer 2 with spatial planning, and layer 3 with disaster management to limit damage following a flood.

Richard Jorissen: 'In the Netherlands we are very good at layer 1, dike reinforcement. If you don't do attend to this, it goes without saying that you'll become very good at evacuation. In layers 2, spatial planning, and 3, contingency planning, we are relatively weaker, whereas there are many opportunities there. How can we make intelligent combinations in the spatial planning of the Netherlands between the state, water boards, provinces, municipalities and safety regions?' Just imagine: proceeding on the basis of multi-layer flood management opens up a world in which it is possible to formulate a planning vision the Netherlands, one that is much broader than only flood protection. Delta Commissioner Wim Kuijken formulates the new Delta Plan as the need to adapt the spatial planning of built-up areas, making the extension from water to space even more urgent. The water sector is potentially the modern version of the defunct Rijksplanologische Dienst: it considers how cities and rural areas can adapt spatially to climate change. But the task must also be seen in a wider context: how do we keep the Randstand future-proof, knowing that this economic hub in north–west Europe is facing huge housing construction and mobility challenges? And, in the rural areas: where is it sensible to construct new houses, and where isn't it? Ingwer de Boer, the recently retired programme director of Room for the River: 'The time has now come for governments to jointly focus on layer 2, spatial planning, so we make sure the risk of flood damage does not increase in the future. Using technologies and design insights developed in civil engineering, urban design and landscape architecture intelligently can produce a cohesive design for the Netherlands.' Centralized control when choosing new residential and commercial areas is not only important to literally keep our feet dry but also to ensure our economic attractiveness is not eroded. This is where the task lies to shape the Delta Plan Flood Risk Management, in the spirit of the Delta Works, into a new creative synthesis and literally extend it beyond the dikes.

Planetary Stewardship

Beyond the dikes lies a whole world. As Jan Terlouw says, 'How do we move from treating symptoms to prevention?' In other words, adapting to the climate is essential but it is completely ineffective as long as we do not tackle climate change. Because that is where the real challenge lies; in that faraway, almost incomprehensible future where the continued existence of humankind is ultimately at stake. For several years the term Anthropocene has been used to refer to our age as one in which human activity has been the most important factor in worldwide change. Water systems are a pre-condition for mankind to live and survive; the task is not only to protect them but also to ensure there is enough fresh water for the world's population. Recognizing that the atmosphere, the oceans and the ecosystems on land influence each other opens the window to understanding that the world needs a more integrated approach in which several issues are investigated simultaneously. 'Planetary stewardship' focuses on the damage resulting from human intervention on earth and searches for milder, more responsible ways to keep the earth habitable and healthy. We are already familiar with the use of natural processes in protecting water systems. The sand motor in front of the Dutch coast has been a ground for a new, sustainable approach to coastal maintenance. A peninsula was created using testing a huge deposit of sand that is being dispersed by the wind, waves and current along the coast, thus creating a stronger coastline. And there are more places where humans have installed 'prostheses' in order to change natural conditions. Man-made islands such as the Markerwadden are an attempt to enrich the ecologically depleted Markermeer. It is hoped that spawning grounds and natural banks will attract plants and animals, thus enhancing biodiversity. A Blue Energy power plant is operating on the Afsluitdijk that produces energy from the difference in salinity concentrations between salt and fresh water. That is water in the Netherlands: we have subjugated it, turned it into land, we swim in it and we skate on it, and now we use it as a source of energy – all that in a land that is as flat as a pancake.

With the Delta Programme, the various implementation programmes of which are scheduled annually, the Netherlands will be prepared until 2050 for the water risks that we can identify at the moment. But new challenges for the years after that are already on the horizon; the water sector will need to quickly develop new solutions. The way technique, technology and nature merge seems to open many opportunities and can even play a role in the multifunctional use of humanly directed natural systems. The technology theoretician Timothy Kaufman-Osborn illustrates this using the reciprocity between a spider and its web: the web is used to catch food and water but it also protects, while the spider continuously adapts it to the changing environment. It is fascinating to contemplate the opportunities that exist and the landscapes that might be created . Globally, there is an opportunity and even, in many cases, a need to revise the way the earth is organized based on water. That task lies far beyond the dikes.

The Zandmotor ('sand motor') along the South-Holland coast, the testing ground for a new, sustainable method of coastal maintenance since 2011

Innovation

Today's hydraulic works are part of a centuries-old tradition of working on flood defences. Innovations are taking place continuously: things can always be done faster, stronger, better or cheaper. Revolutionary inventions are rare. The tradition is one of continuity and gradual change. The flood defence programmes of recent decades demonstrate substantial innovations in processes (How do you tackle that?) and modest innovations in technology (What are we going to make?). At a time of large projects, complex tendering processes, and a legalistic, decentralized and privatized approach, the challenge is to create room for renewal and unconventional ideas. This includes the inevitable failures. As part of the Room for the River programme a brochure was published called 'Ruimte voor innovaties, stapsgewijs of met grote sprongen' ('Room for innovations, step by step or in great leaps'). It argues that the power of innovation is greatest when an urgent problem needs to be solved. The urgency of recent flood defence works was urgent on two fronts: the search for an alternative to mere dike reinforcement, and speeding up processes.

Moving with the Rivers: Innovation?

The step from defensively reinforcing and raising dikes to the more offensive moving with the water created a complete turnaround in thinking about flood defence in the Netherlands. But was it also a revolutionary innovation? Moving with the water is not a matter of choice in many parts of the world: it is a bitter necessity. Nature development and secondary channels were already being used along the rivers of the Brazilian city of Curitiba in the 1970s (see p. 210). In the twentieth century flood defence in the Netherlands was always about spectacular civil and hydraulic engineering works, climaxing in the impressive Delta Works. The idea was that a delta could be engineered into conformity, that it could be shaped on the drawing boards and that, through technology alone, the dangers could be banished. The transition from a technocratic to a more organic approach to the water challenge was a long time in the making. After the near-disasters of 1993 and 1995 this process suddenly accelerated and began to be implemented. The revolution lay not in spectacle this time but in a rediscovered naturalness combined with a hi-tech flood defence. The solution to the flood defence question was found in the concept of a natural system – a concept that had been dismissed since 1900 precisely because it was supposedly dangerous and replaced with all sorts of man-made interventions.

Retro Innovations

The concept of giving rivers more room again led to various tried and true innovations being recycled, the return of the *terp* (artificial mound) being just one example. For centuries these mounds protected inhabitants in the north of the country; now this high-and-dry solution to living in flood-prone areas was rediscovered by the inhabitants of the southern Overdiepse Polder (see p. 132), and it was also implemented in the Noordwaard (see p. 140) and Voorsterklei. *Terps* and dikes were used in other projects to create shelters for animals during high water (see p. 116). Returning dunes to the Hondsbossche Sea Defence also marked the renewed use of a traditional flood defence (see p. 220). And what about an innovation that discards the complex and expensive automatic operations of hydraulic systems, relying instead on a a tractor with a battery or a crane or. The battery operates the pneumatic *melkmeisjes* ('milkmaids') on the lifting gates of the hypermodern inlet structure of the high water channel in Veessen (see p. 126). A crane is also used to lift the stop logs into place at the equally innovative spillways in Pannerden and the Hondsbroeksche Pleij (see p. 120). Both examples demonstrate a no-nonsense approach, focused on cost efficiency, operational reliability and an extremely long service life. Low tech is the ideal solution for operating systems that will be used once in 10, 20 or 80 years – and when there will be no room for failure.

Invisible Innovations

A second type of innovation is concealed in existing dikes, dunes and barriers. These are upgrades of existing systems that are made stronger, more climate-proof, more predictable, and sometimes oversized as a result. The coast has been invisibly enriched with a dike-in-dune and a dike-in-boulevard, for example (seep. 94, 100). Cavities built into stone revetments turned these coverings into wave attenuators and also encouraged biodiversity (see p. 66). The sea defence in Harlingen can bend with the storm rather like a reed, albeit it a slightly stiff one, having been fitted with a concrete flood defence on a foundation of swaying tension anchor piles, each 40 metres long (see p. 188). In the river dikes indiscernible innovations have been implemented to counter piping (when sand is washed out of dikes) and seepage (rising groundwater). Augmented reality has also been used in design processes. Smart sheet pile walls will subside with future soil subsidence (see p. 196), and new dike houses can be jacked up to keep pace with dike raising (zie p. 74). It is still unclear how enduring new solutions such as geotextile, dike nails, dike dowels, open-cell concrete revetments and super glue will be in the long run. This is all part of experimentation and requires research and monitoring. Here, too, high tech makes its appearance, with measuring apparatus that is built into the dikes, and drones or jet fighters that carry out inspection flights. Conventional test methods continue to be used alongside these more high-tech approaches. Allowing the controlled collapse of a traditional dike and one reinforced with sheet pile walls is just one example. Tests like this produce a wealth of information about the behaviour of dikes under intense pressure.

Visible Innovations

Fifty years ago the Netherlands was hard at work on the Delta Works. Entire sea arms were closed off, resulting in spectacular engineering works. The power-play of engineers can still be discerned in the storm surge barriers, locks and advanced dams, exuding as they do monumentality, ingenuity and advanced technology. By comparison, recent hydraulic works are modest: blended into the landscape and discreet. The dominant impression created by these works is that of an apparently untouched nature, comparable with the sea and river views of about 1900. Only

The openings in the Oosterscheldekering made it possible to retain the salt water and tides in the Oosterschelde

Typically Dutch retro innovation: living on a terp in the Noordwaard

occasionally is a modern accent detectable. The bridges in Nijmegen, for example (see p. 236), or the inlet in Veessen, an impressive gallery for the water, hidden under a bridge by the meadows (see p. 126). The spillways at the bifurcation of the Waal and IJssel rivers: large concrete pergolas in the floodplains (see p. 120). And the wave-attenuating dike at Fort Steurgat in the Noordwaard, a stand of willows that acts a wave-breaker. This made it possible to keep the dike here substantially lower than planned and still ensure that it met flood defence norms (see p. 140). One of the most visually striking and at the same time best concealed innovations is in the harbour of Spakenburg: a flood defence that automatically inflates and floats itself (see p. 206).

Process Acceleration as Innovation

New legislation has made it possible to save considerable time during the design, permit, tendering and implementation processes. An interesting innovation can be seen in the collaboration between national and regional governments. The national government establishes the pre-conditions (aim, money, time) for projects, but regional governments and authorities (provinces, municipalities, water boards) are given the room to shape projects as they see fit. They work together with residents and business, which creates wider support, reduces delays and sometimes produces better ideas. To keep projects firmly in hand, detailed arrangements are put in place concerning content and process, risk management, and interim plan changes. The national government provides experience via a programme office and expertise clusters (spatial quality, legal affairs, soil, river engineering and integral design). And regional governments are given the opportunity to advise on the formation of plans and to direct the detailing and implementation of those plans. Collaboration is aimed at reaching a consensus and combining flood defence with other spatial assignments. This set-up includes speeding up processes wherever possible, for example, dealing with a number of steps simultaneously: accelerating and expediting permit procedures, appeals, tendering processes, land acquisition, as well as involving managers and dealing with claims for damages. The process is strictly managed to ensure budget and planning goals are met, despite the complex assignments, the number of parties involved and the many contingencies. It is remarkable, given the Netherlands' tradition of vast civil engineering works, how many projects were completed ahead of schedule and within budget.

The inlet structure of the IJssel high water channel at Veessen

Thanks to the dike-in-dune solution it wasn't necessary to raise the dunes in the centre of Katwijk and the view of the sea was preserved

Variations on a Dike or Dune

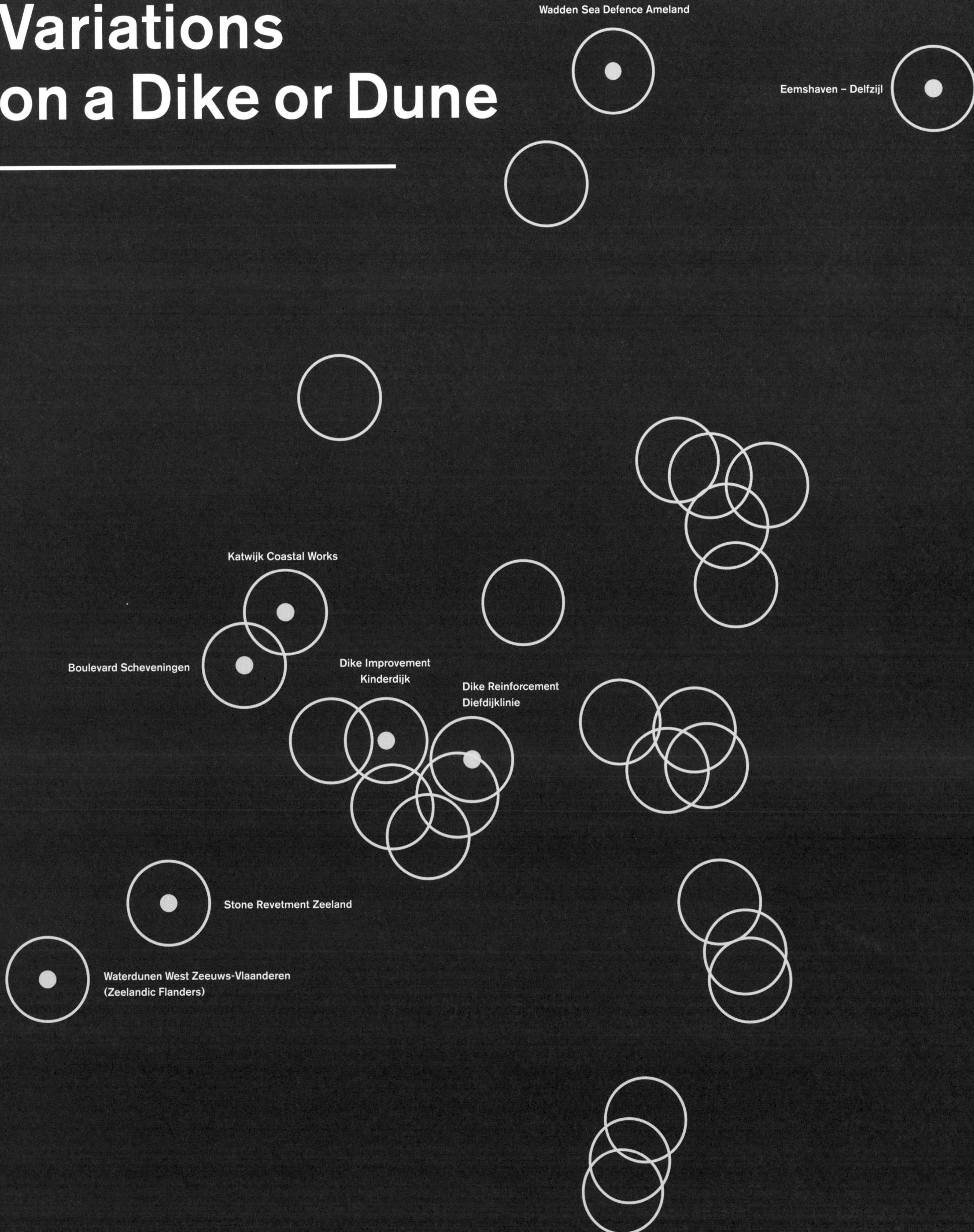

Wadden Sea Defence Ameland – The Wadden Sea attracts children to play along the dike in the Ballumerbocht, 2017

Stone Revetment Zeeland – The dike being reinforced along the salt marshes of the Verdronken Land in Zuid-Beveland, 2009

Dike Improvement Kinderdijk – Schoonhovenseveer – The robust Klimaatdijk (climate dike) at Streefkerk, 2017

Sample Sheet of Intervention – overview on page 24-27

Breakwater and habitat for flora and fauna

Programme: **independent project**
Province: **Zeeland**
Coordinates: **51°36'12.3"N 3°49'00.8"E**
GPS: **51.603420, 3.816899**
Assignment: **coastal reinforcement, primary flood defences**
Implementation: **1997–2015**
Water board: **Waterschap Scheldestromen**
Design: **Projectbureau Zeeweringen together with consulting engineers and freelancers**
Contractor: **Boskalis, van Oord, KWS, Heijmans, Liebregts, van den Biggelaar, van den Herik, Hakkers et al.**
Length: **325 km dike**
Other figures: **607,000 m² tilted concrete blocks; 2,860,000 m² concrete columns; 1,519,000 m² rubble and liquid asphalt; 93 km new cycling paths across maintenance paths**

The Schelde (Scheldt) river flows from Belgium into the Netherlands, and once flowed into the North Sea through two sea arms, the Oosterschelde and the Westerschelde. Nowadays all its water is discharged via the Westerschelde, which has an unimpeded opening to the sea and forms the entrance route to the port of Antwerp. The Oosterschelde has not been fed by the Schelde since the nineteenth century. In 1986, as part of the Delta Works, this sea arm was closed off by a storm surge barrier consisting of 65 piers and 62 enormous, movable sluice gates that can be lowered if there is a strong storm. The ebb and flow of the tide and the salt-water environment behind the dam were retained. The Oosterschelde is the largest national park in the Netherlands. The dikes along both sea arms have been reinforced in recent years.

Decisive Leadership

The alarm was sounded in 1996 by the Waterschap Zeeuwse Eilanden water board, which became part of Waterschap Scheldestromen in 2011: inspections by Rijkswaterstaat had shown that at several points the Zeeland dikes were in a poor

Completed dike reinforcement along the Westerschelde at Ellewoutsdijk

state. The stone revetments were too light. During a storm with powerful wave action, concrete blocks might collapse as a result of the clay layer underneath them being washed away. The blocks and basalt boulders that were fixed in place might also give way. The Minister of Transport, Public Works and Water Management Annemarie Jorritsma immediately ordered a thorough inspection. As early as 2 April 1997, precisely one day after the storm season had ended, she gave the starting signal on the dike at Borssele launching the Zeeweringen (sea defences) project. The *Provinciale Zeeuwse Courant* newspaper complimented the authorities, saying 'Swift political action is possible when it comes to the crunch.'

Time-Consuming Puzzle

Despite the resolute decision-making, it still took 18 years before the dikes along the Ooster- and Westerschelde had been given a heavier stone revetment. As a Dutch saying has it, most time is lost in preparatory work ('met passen en meten wordt de meeste tijd versleten'). The Zeeland dikes were reinforced using nearly 5 million m^2 of tilted, reused concrete blocks, new concrete columns, and rubble and liquid asphalt. The choice of revetment had to be determined separately for each of the 108 dike sections, whose total length is 325 kilometres.

Nature and Environment Both Decisive Factors for Choice of Material

The 1996 Wet op de waterkering (Flood Defences Act) applies to the reinforcement of dikes. But the waters of Zeeland are also of high natural value and they are protected by the Dutch Flora- en Faunawet (Flora and Fauna Act) and by European legislation. Nature and the environment were important criteria when selecting revetment material. The dike authority inventoried the salt vegetation and seaweeds for each dike section that was to be upgraded. Material for the dike revetment was then chosen that enabled the affected vegetation to return or, better still, stimulated it to develop further. Sea grass was also moved here and there, bee hotels built for the rare sea aster mining bee, and tidal pools created for fish, sponges, crabs and shrimps. Thanks in part to the efforts of researchers, manufacturers and construction companies, every opportunity was taken to realize the project as sustainably as possible. Take, for example, the concrete columns (hydroblocks): giving them a 'waistline' reduced the material required by 40 per cent. And because more of these lightweights fitted into a single lorry, fewer transport movements were required.

Rougher Surface Benefits Seaweeds and Sea Creatures

The interaction between engineers and biologists resulted in the use of 'soft' stones with an open and rough surface, including Basalton and Haringman blocks with an ECO top layer, which plants can easily

Eerste Bathpolder, dike Oosterschelde Applying canvas and rubble to the clay of the dike body, 2009

The slope of the dike is covered with a revetment made of concrete pillars

attach themselves to. The more open materials also break wave action better than a smooth slope with grass or asphalt. The Rotterdam manufacturer Hillblock developed dike stones that retain more water thanks to their rougher structure. These stones provide a nourishing bed for seaweeds and sea creatures in the tidal zone, the zone between ebb and flow. After only two months the first periwinkles, barnacles and brown algae such as wrack, bladder wrack and knotted kelp were growing here. Another solution was found for the rubble. A layer of lava stone was scattered over the waterproof liquid asphalt that was used to fixate the rubble. This created a porous top layer that also encouraged the attachment of seaweeds. Several other dikes along the Dutch coast have been covered with materials that are hospitable to flora and fauna. In the north of the country they are referred to as *rijke dijken* (rich dikes) (see p. 70, 90).

Retaining Cultural History

Various sites of cultural and historic value were included in the dike design. The historic Ellewoutsdijk fort, for example, was saved by reinforcing the two dikes on either side of it. Old tidal harbours such as the Flaauwershaven in Kerkwerve were preserved wherever possible and renovated. On Noord-Beveland the three *nollen* – remains from old dike breaches – were reinforced from behind, thus preserving the original revetment. At the Plompe Toren near Burghsluis and elsewhere *Muralt* walls were retained. These small one-metre-high walls made of concrete slabs were added to many Zeeland dikes after a disastrous 1906 storm. Though cheap and flimsy, the wall remains are listed *Rijksmonumenten* (national heritage sites). The new stone revetment also had to suit the Zeeland landscape. Now that this Zeeland flood defence project – the largest since the Delta Works – has been completed, the people of Zeeland can rest assured that their dikes are safer than ever. All this beauty can best be explored on a bicycle. The total length of outer-dike maintenance paths that also function as cycle paths has been doubled to some 200 kilometres.

New stone revetment on the Oosterscheldedijk; the Zeeland bridge is on the horizon

Old *Muralt* wall on the reinforced dike in Burghsluis; the Plompe Toren is in the background

Sample Sheet of Intervention – overview on page 24-27

A Logistics Puzzle

Programme: **Second Hoogwaterbeschermingsprogramma (Flood Protection Programme)**
Province: **Friesland**
Coordinates: **53°26'36.3"N 5°43'54.9"E**
GPS: **53.443415, 5.731913**
Assignment: **reinforcement Wadden Sea dike**
Implementation: **2014–2018**
Water board: **Wetterskip Fryslân**
Designer(s):
Contractor/consortium: **consortium Van den Herik Sliedrecht and Jansma Drachten**
Stakeholders: **Gemeente Ameland, Provincie Fryslân**
Length: **16.5 km**
Earthworks: **365,000 tonnes rubble; 150,000 tonnes asphalt; 235,000 tonnes clay; 50,000 tonnes sand**
Other figures: **dike heighted 35 cm on average**

The north coast of the Netherlands is bounded by a series of islands called the Waddeneilanden (Wadden islands). They border the Wadden Sea, a UNESCO World Heritage site that extends all the way to Germany and Denmark. It is a cultural landscape with great natural values, partly because most of the Wadden Sea is uncovered during low tide. The Wadden islands, Texel, Vlieland, Terschelling, Ameland and Schiermonnikoog, have a long history of seafaring and agriculture. For decades, tourism has flourished here thanks to the well-organized ferry service, nature, and the many beaches, cycle paths, holiday homes and hotels. All of the islands have a similar elongated shape. The north side consists of beaches and high dunes that provide protection against the North Sea. On the south side polders and dikes form the border with the Wadden Sea. Within the frame-work of the Hoogwaterbeschermingsprogramma (Flood Protection Programme), many dikes in the Netherlands are being raised or reinforced. Ameland has a 16.5-kilometre-long sea dike on a coastline that extends for more than 36 kilometres. The reinforcement and raising of this dike was the largest dike project ever undertaken by Wetterskip Fryslân.

Raise and Reinforce

Reinforcing the dike on Ameland was not particularly complex. The existing dike failed to meet the required standards in 2006 because the revetment of grass and concrete blocks on its outer side was no longer adequate. The reinforcement comprised several steps. The bottom of the dike on the sea side, its 'toe', was renewed with canvas and a layer of rubble. The

Ameland's two coasts are clearly visible: the North Sea beach on the right and the Wadden Sea dike on the left

The sand required was shipped to Ameland and distributed from there

Step-shaped seating has been built into parts of the sea flood defence

stones in the rubble were different sizes and types, making it easier for flora and fauna to establish themselves. This is known as the *rijke dijk* (rich dike) concept (see p. 90). The dike's toe was bordered by an asphalt cycle path, which is now 90 centimetres higher as a result of the intervention. The top of the dike was filled with clay, covered with geotextile, asphalt, debris (mixed granulate), clay and grass. Applying fibrous open stone asphalt meant the grass could set root through the asphalt. The renovated dike has been raised by on average 35 centimetres, but its exterior is unchanged.

Taking All Seasons into Account

The greatest challenge when working on the Wadden islands is planning. Winter brings the wild storm season. Disturbing bird colonies during the breeding season is not permitted, and the birds' flood shelters are also out of bounds. During the tourist season in the summer the islands are crowded and it is unacceptable for visitors to be inconvenienced by construction works. However, most of the tourists pay no attention to the Wadden dike, hurrying instead from the ferry to the dunes and the villages. And so it was decided to carry out the work between April and October after all. Disruption was minimal because only certain dike sections were tackled at any one time. Transporting 800,000 tonnes of rubble, clay, sand and asphalt to Ameland was no mean task. Two coasters carried all the material from Harlingen to three different unloading bays on the island.

Taking Humans and Animals into Account

Ameland has an abundance of natural life, both on and around the island. Its mudflats and dunes are a protected Natura 2000 area. The inner-dike area is also part of the National Ecological Network. At every stage in the process, natural values had to be taken into account. On the dike itself, 19 eight-metre-wide crossing points were made for ducklings because they could not get across the debris revetment. At the Feugelpôlle bird breeding grounds outside the dike, clay was smeared over the rubble to cover it so that young terns and black-headed gulls would not get stuck in it and die. The dike reinforcement turned into an excursion for tourists, with the 'Kijk op de Dijk' ('view of the dike') exhibition and a Wadden balcony providing an overview of the Wadden Sea and the dike works. Tourists can cycle outside the dike along the Wadden Sea. Steps large enough to sit on have been built into the dike at the Ballumerbocht and the Burensteiger. Kite-surfers, just like the ducklings, terns and black-headed gulls, were given their own concrete-block crossing point through the rubble.

The dike's outline can clearly be experienced in the Ballumerbocht

Bird-watching at sunset at the high water shelters

One of the delivery depots for the Nes port project

02
03
04
Sample Sheet of Intervention – overview on page 24-27
10
31

Variations on dike reinforcement

Programme: **Second Hoogwaterbeschermingsprogramma (Flood Protection Programme)**
Province: **South Holland**
Coordinates: **51° 56' 20.158" N ; 04° 50' 34.83" O**
GPS: **51.939554**
Assignment: **dike reinforcement**
Implementation: **2013–2017**
Water board: **Waterschap Rivierenland**
Design: **Van der Padt Architecten en Partners**
Contractor: **Combinatie Dijkverbetering Molenwaard (Mourik and Besix)**
Management: **Waterschap Rivierenland**
Length: **17-km dike, of which 11 km improved**

If 450 houses and many commercial premises are located on and next to a river dike, dike reinforcement can look like an impossible task. And it looks even more hopeless if some of these buildings have come to lie deeper in the dike as a result of earlier dike reinforcements. But the Waterschap Rivierenland water board had no choice regarding the 17-kilometre-long Lekdijk between Kinderdijk and Schoonhovenseveer. More than half of the dike did not meet the legally prescribed norms. By using and combining all sorts of innovative techniques, the task was successfully carried out with minimal damage and inconvenience. Only 58 houses had to be demolished, and even this was an interim measure.

Four Different Measures

Four construction measures were applied across the dike trajectory, depending on what was possible and necessary. First, where there was enough room inside the dike it was raised and reinforced with soil. Second, pile and diaphragm walls were used for the houses. This was the first dike where anchored bored piles with a one-metre cross-section (1,100 in total) were used. Because the bored piles were installed using minimum vibration and no soil displacement – i.e., without pushing the soil away during ramming operations – the houses suffered minimal to no damage. Third, slender diaphragm walls were guided into the ground at Streefkerk and Nieuw-Lekkerland. Fourth, the dike was repositioned in three

In the background the slope of the new Klimaatdijk (climate dike); the dike road is embedded anew next to the crest

places on the outer river side to improve the stability of the existing dike.

Monitoring

During the construction works the effects on the substrate and the buildings were monitored in all sorts of ways. Each house was fitted with a vibration meter. Settlement rods, which are used to assess the subsidence of a dike body, were surveyed using a drone. This drone required only seven minutes to survey a whole hectare. When the survey had been completed, all the data was made available and usable employing a 3-D scan. All the points that are traditionally surveyed by a land surveyor were covered. It was also possible to read out the height every 2 centimetres. Dike repositioning was monitored using a fibreglass cable that was installed in the ground specially for this purpose. Moving the substrate near houses was also scrupulously monitored using wireless sensors. All the parties that participated in the dike reinforcement project, from designers and those involved in preliminary work to contractors and the client, had access to the same digital source of information. Via BIM – a 3-D building information model – all of the project information about houses, roads, cables, lines, cycle paths, vegetation, heights and the like was collected, linked and updated. In a nutshell: the right information at the right moment for the right person. Augmented reality was also used to give residents an idea of what their surroundings would eventually look like.

Climate Dike

A revolutionary dike reinforcement technique was used at the marina in Streefkerk: a multifunctional climate, or Delta, dike. Such a huge amount of soil was used here that the dike is 100 times stronger than prescribed by the current safety norm. It can withstand climate change and rising water levels for much longer than the legally prescribed period. Besides this, there is room for housing on the dike and recreational development of the easily accessed river bank. A similar oversized dike was also constructed in the Munnikenland: the Wakkere Dijk (see p. 136).

Houses That Can Be Jacked Up

The dike had to be reinforced using a shoulder in a number of places. In order to create room, 58 houses had to be demolished but half of the residents returned. An agreement had been reached allowing them to build new houses on the shoulder when the project had been completed – houses that could be jacked up. When the dike needs to be reinforced in the future, the demolition ball will not have to be used again because the houses can be raised intact. They have strong foundations and extendable steel poles covered by a concrete plate. The plate is 'simply' jacked up, house and all, and the dike-improvers can set to work.

A total of 1,100 bored piles keep the dike in place where there is not enough room for widening the sand body

The bored piles are linked to each other and anchored with a concrete cap

A shoulder with a gentle slope was added to the Lekdijk at Groot-Ammers

SCHEEPSWERF
GROOT. AMMERS.BV
ARRIVA

Sample Sheet of Intervention – overview on page 24-27

A defence in every sense of the word

Programme: **Hoogwaterbeschermingsprogramma (Flood Protection Programme)**
Province: **South Holland and Gelderland**
Coordinates: **51°56'07.9"N 5°09'09.2"E**
GPS: **51.935534, 5.152561**
Assignment: **dike reinforcement**
Implementation: **2014–2017**
Water board: **Waterschap Rivierenland**
Design: **AnteaGroup**
Contractor: **Diefdijkalliantie (G. Van der Ven BV Aannemingsbedrijf and Sterk Heiwerken BV) J. den Boer BV, Ballast Nedam**
Project Manager: **Waterschap Rivierenland**
Length: **23 km dike, of which 8 km reinforced**
Other figures: **41,520 m^3 berm and earthworks; 475 m seepage screens; 1,455 m stabilizing diaphragm walls; restore 18 historic objects along Nieuwe Hollandse Waterlinie; 74,000 m^2 road maintenance**

The ancient Diefdijk lies on the border between the provinces of Gelderland and South Holland. It connects the Lek river at Everdingen with the Merwede river at Gorinchem. The high, steep dike dates from 1277. The population was increasing rapidly at the time, creating great demand for farmland. Every time a river dike was breached in Gelderland, the Holland polders to the west were flooded and harvests were ruined. The Diefdijk was intended to prevent this. Through the centuries, the dike was reinforced and raised using soil and vegetative material. This practice weakened the dam at various points, as shown during inspections carried out within the framework of the Hoogwaterbeschermingsprogramma (Flood Protection Programme). There was also a danger of 'piping', when water seeps through the dike and takes sand with it, causing the substrate to collapse and weakening the dike. The necessary dike reinforcement was linked to improving road safety and opening up the dike's rich military history. Experts were linked to the project's aims using a single plan, a single government agency, a single legal procedure and a single contractor.

No Flood Risk in Alblasserwaard and Vijfheerenlanden

Being part of the coherent system of dikes known as the Diefdijklinie, the

The dike along the Linge has been reinforced with rubble on a bed made of plaited willow branches

Dike reinforcement next to historic buildings is carried out with particular care

Places along the Diefdijk with too little space were reinforced with sheet pile walls

Diefdijk has retained its flood protection function throughout the centuries. If one of the dikes in the Betuwe or the Tieler- and Culemborgerwaarden collapses, the water will not pass this line. The lower-lying Alblasserwaard and the Vijfheerenlanden are thus protected against flooding. The stored water can flow away via overflows in the river dike near Gorinchem as soon as the water level in the Merwede has dropped enough. The Diefdijklinie has several *coupures*, corridors that can be closed off quickly in an emergency. One of them is the viaduct over the A2 motorway in the Diefdijk, slightly west of Culemborg. The motorway can be closed off with sliding gates in the event of high water.

New Clay Layer, Dam Walls, Seepage Screens and Rubble Layer

Reinforcement measures have been taken at various points on an 8-kilometre stretch of the dike. Improvements have been implemented in the ground wherever possible. Where there was enough room, a new clay covering was applied. The existing top layer was removed first to ensure that this clay layer would stick to the old layer properly. Sheet pile walls or seepage screens were constructed in places where houses or natural and cultural-historical values had to be taken into account. Along the Linge rock fill was also used, thus reinforcing the underside of the dike with stones.

Retention of Characteristic Landscape

The area's residents and the Projectbureau Nieuwe Hollandse Waterlinie (see text box) played a crucial role in the design. One of the aims was to retain the beauty of the landscape. The characteristic dike profile, the pretty farms and the tall standard fruit trees that are so typical of the area were preserved. So, too, were the eddies and *wielen*, deep lakes immediately behind a dike formed when it is breached. The most beautiful of these is the Schoonrewoerdse Wiel or Wiel van Bassa, which was formed in 1573. With a surface area of 13 hectares and a depth of 8 metres, it is the largest *wiel* in the Netherlands.

Drive Slower and See More

Road safety was an important issue. People always drove quite fast on the Diefdijk, a dangerous practice, especially at crossings where differences in height reduce visibility. New traffic-calming devices have slowed the traffic and discouraged motorists using the dike as a short-cut. Some of these devices have been installed on both sides of the dike near objects of historic interest, subtly drawing attention to them. Support for the interventions was widespread because all the parties worked together. The works were implemented coherently, saving a great deal of time and money.

The Diefdijk crosses the A2 Motorway; the coupure can be closed off when water levels are high using sluice gates (see p. 45)

The Diefdijk in all its glory, a characteristically winding, narrow ribbon

Nieuwe Hollandse Waterlinie

The Diefdijk was at first only intended to prevent flooding. But in the eighteenth century it was also given a military function, first as part of the Hollandse Waterlinie and, from 1815, as part of the Nieuwe Hollandse Waterlinie. These were defensive lines based on the inundation principle: making the lines impenetrable by inundating lands until the water was knee-high. The Nieuwe Hollandse Waterlinie is 85 kilometres long, effectively cutting the west of the Netherlands off from the outside world. It extends from the island of Pampus near Amsterdam to the Biesbosch. Besides inundation areas, the line comprises forts, fortified cities, locks and bunkers. The Nieuwe Hollandse Waterlinie has been partly or completely inundated three times: in 1870 during the Franco-Prussian War, and in the First and Second World Wars. Famous sites along the Diefdijk include Fort Everdingen and Fort Asperen. When the Diefdijk was reinforced, trenches, gun emplacements and bunkers were restored to their old glory and made accessible via footpaths and dike steps.

The Nieuwe Hollandse Waterlinie cultural heritage site is acquiring a new educational and recreational function. An old trench has been restored next to this casemate and gun turret near Acquoy

'Bunker 599' is a project by landscape architect Ronald Rietveld and Atelire de Lyon. The sliced-through bunker and the bridge that disappears under water provide a good insight into how the Nieuwe Hollandse Waterlinie worked

Sample Sheet of Intervention – overview on page 24-27

Super-strong link integrates flood protection, nature, economy and quality of life

Programme: **Second Hoogwaterbeschermings-programma (Flood Protection System)**
Province: **Zeeland**
Coordinates: **51°24'23.3"N 3°30'42.0"E**
GPS: **51.406475, 3.511675**
Implementation: **2013–2015**
Water board: **Waterschap Scheldestromen**
Design: **Buro Ruimte & Groen, Borssele (fringe zone)**
Contractor: **De Vries & Van de Wiel Kust- en oeverwerken, Heijmans Wegen**
Project Manager: **Het Zeeuwse Landschap, partly in collaboration wtih Molecaten**
Surface: **350 ha, of which 14 ha dune camping, 40 ha holiday bungalows/hotel, 250 ha nature (including 173 ha inner-dike salty tidal nature)**
Earthworks: **1.5 million m³ soil and 400,000 m³ sand**

The periodic inspection of Dutch flood protection systems in 2006 showed that the coast of west Zeeuws-Vlaanderen (Zeelandic Flanders) was a weak link. This is the most southerly point of the Dutch coast and borders on Belgium. A single 'super storm', with a risk of 1 in 4,000 per year, would be catastrophic. The weak points along this coast were addressed through five sub-projects: Cadzand-Bad, the rediked Zwarte Polder, Nieuwvliet-Groede, Waterdunen and Breskens. A tailor-made solution was sought for each area. New dunes were constructed in several areas and the beach was shifted some distance into the sea. This was made possible by sand supplementation: spouting out large quantities of sand onto the beach. This was done from the sea using trailing suction hopper dredgers. It was not possible to reinforce the sea side at Waterdunen, between Breskens and Groede, because the strong current in the deep tidal channel here would have returned the sand to the sea in no time. An inland solution was found, one that will protect the land directly behind the dike against rising sea levels for two centuries rather than only the standard 50 years. The weak link was transformed into a super-strong one.

Salty Nature and a High-Quality Camping Site

It all started with two private organizations. The nature conservation body Het Zeeuwse Landschap had already been looking for a location for large-scale salty nature development and an international reserve for migratory birds. Molecaten, an important player in the leisure industry, owned a camping site along the coast near Breskens that it wanted to upgrade.

The old situation shows the former camping site behind the dike, 2011

A new water landscape is created. The inner slope of the dike has been reinforced by constructing a new dune, 2014

Reinforcing the sea dike at the Breskens lighthouse, 2015

24
27

The rough outline of the new Waterdunen landscape is already visible, 2016

Molecaten and Het Zeeuwse Landschap entered discussions and came up with a plan to combine coastal reinforcement with their own goals. The state, Province of Zeeland, Scheldestromen water board and the municipality of Sluis considered it an excellent proposal. Joining forces, they started to develop the 350-hectare Waterdunen nature and recreation area in the Oud- and Jong-Breskenspolder in 2013.

Climate Dune Behind the Dike

Inland, a new dune was developed against the existing dike using clay from the polders behind it – 1.5 million m^3 of soil and 400,000 m^3 of sand. Digging out creeks and channels in the area provided plenty of construction material. The sand was transported from the North Sea using sand dredgers and piped into a sand depot. The climate dune is 300 metres wide and has an average height of 14 metres. The highest dune is 18 metres high and sticks out above the crest of the dike by 7 metres. The sea side of the dike therefore only had to be upgraded with a stone revetment and asphalt.

Tidal Culvert Brings Salty Westerscheldewater into Nature

A second novelty is the tidal culvert in the dike part of the Westerschelde. Four concrete cylinders with an interior measurement of 4 x 3 metres run straight through the sea defence. Salt water from the Westerschelde will flow into the nature area via three of them from 2017-2018. When the tide is low the water will retreat. The fourth cylinder is intended for draining excess polder water in the nature area. The inflowing sea water will be regulated using movable steel gates. These can be operated using remote controls and sealed hermetically during a storm surge. The water flowing in and out will be put to good use: a tidal power station has been built into the culvert to generate clean electricity for the Waterdunen recreation area. This set-up resembles the Dubbele Dijk, which is being constructed at the other end of the Dutch coast, on the eastern edge of Groningen (see p. 90).

Unique Delta Nature for Birds and Aquaculture

The tidal culvert makes it possible for an international hotspot for migratory birds and unique delta nature to develop in Waterdunen. The creeks, salt marshes

The large new longitudinal groyne in the beach at the Herdijkte Zwarte Polder

and mud flats are a plentiful source of nourishment: insects for swallows and meadow pipits, grass and aquatic plants for geese and ducks, and shellfish and worms for waders. Observation huts and screens as well as a bird-counting post have been built for bird-watchers. Het Zeeuwse Landschap manages this natural reserve from an old farm at its heart. The Kustlaboratorium voor aquacultuur, a coastal laboratory for aquaculture, will also be built here. Het Zeeuwse Landschap plans to grow salt vegetables here and farm shellfish to show that aquaculture inside the dike fits in perfectly with the natural surroundings. A fresh-water seepage ditch has been dug around the salty nature area to prevent salt water flowing into the surroundings and to drain polder water into the sea.

Economic Impulse for the Region

Molecaten has moved its operations to the dune area. A dune camping-site, 400 bungalows and a hotel with 80 beds have been developed in a manner that accentuates different types of landscape. Nature here is managed jointly by Molecaten and Het Zeeuwse Landschap. The 55-hectare Natuurvakantiepark Waterdunen also hosts a boulevard with small-scale hospitality and retail operations. Cycling, hiking and bridle paths have been constructed throughout the area. The extensive facilities and the special interaction of water and land, salt and fresh water, ebb and flow, attract tourists and recreational users to Waterdunen throughout the year. The resulting economic impulse creates jobs and improves the quality of life, both badly needed in the region. Based on economic and spatial developments, the tide also has to be turned on these fronts, something the regional partners address in their 2020 Gebiedsplan Natuurlijk Vitaal plan.

The new marina at Cadzand-Bad, hugged by protective break-waters

02 03 04 Sample Sheet of Intervention – overview on page 24-27

Earthquake-proof dike

Programme: **Second Hoogwaterbeschermings-programma (Flood Protection Programme)**
Province: **Groningen**
Coordinates: **53°24'05.7"N 6°52'54.4"E**
GPS: **53.401589, 6.881772**
Assignment: **dike reinforcement and earth-quake-proofing**
Implementation: **2014–2019**
Water board: **Waterschap Noorderzijlvest**
Designe: **Combinatie Ommelanderdiek, D&C contract (dike), Bosch Slabbers and Alterra Wageningen UR (double dike)**
Contractor/consortium: **Combinatie Ommelander-diek (Boskalis Nederland and KWS Infra)**
Stakeholders: **Province of Groningen, municipalities of Delfzijl and Eemsmond, Rijkswaterstaat, Stichting Groninger Landschap, Nationaal Coördinator Groningen**
Length: **12 km**

Gas extraction has taken place in the province of Groningen since 1959, making an incalculable contribution to the Netherlands' prosperity. But gas extraction causes damaging land subsidence and earthquakes. These earthquakes also threaten some of Groningen's flood protection systems – a total of 40 kilometres of sea dikes (primary flood defences) and 60 kilometres of canal dikes (regional flood defences). These flood defences must meet safety norms regardless of the earthquakes. 'Softness' in dikes is a common problem during earthquakes: the vibration during an earthquake causes a sort of quicksand to develop under the dike, which then collapses and cracks. No norms exist for such events because the Netherlands has no experience in dealing with flood defences that are exposed to earthquakes; moreover, the chance that an earthquake will coincide with a flood is minimal. Waterschap Noorderzijlvest together with numerous consultants therefore investigated how the necessary reinforcement of the sea dike between Eemshaven and Delfzijl could be made earthquake-proof.

Also Safe During an Earthquake

A 12-kilometre stretch of the sea dike that extends from Eemshaven to Delfzijl did not meet the norms set out in the framework of the Hoogwaterbeschermings-programma (Flood Protection Programme). The dike had to be raised by an average of

The robust outlines of a new stretch of dike at Delfzijl

21 25 27 29 30

A gas-drilling location just behind the Eemsdijk at Bierum

The earthquakes have left their mark, for example in this old building in Zeerijp

50 centimetres, increasing to 2 metres in Delfzijl. The dike was strengthened by widening the land side by 5 to 15 metres. It was also necessary to ascertain whether, in the event of an earthquake, the dike might be so badly damaged that it no longer complied with safety norms. An earthquake with a maximum force of 5 on the Richter scale was assumed, stronger than the most serious quake recorded in Groningen (3.6 in Huizinge in 2012).

Preventing, Monitoring and Repairing

To make the dike earthquake-proof, a vastly expensive solution was first considered: placing the dike body between two sheet pile walls. On the advice of experts in the United States and Japan, the soil under the dike was examined further, and possible liquefaction was investigated using a sophisticated calculation method. This showed that the sand under the dike was mixed with clay and softens more slowly than initially feared. The calculated deformations also turned out to be less serious than first thought. This made the necessary intervention easier and cheaper. A special fabric (geotextile) installed in several places in the dike sufficed, as it would prevent sand being washed out of the dike. Besides this, clay depots were constructed along the dike so that damage can be repaired quickly. The third measure in relation to the earthquakes was the installation of sensors so that movements in the dike can be monitored precisely. NAM, which exploits the gas field, met the costs of these special measures.

A Double Dike

At several places the dike reinforcement project was used to implement new initiatives. In Delfzijl the dike was repositioned slightly to make way for a city beach, for example. A little further along, a double dike is taking shape on 2.5 kilometres of the trajectory, with the construction of a second dike behind the primary flood defence. In the area between, the seawater can flow in through a culvert during high tide, rather like the culverts in Waterdunen in Zeeland (see p. 84). Some 55 hectares of tidal nature will be developed: a paradise for birds where, moreover, the sludge can be harvested. A separate part of this 'in-between' area will be developed for marine farming: growing salt crops such as glasswort and sea lavender, and farming crustaceans and shellfish. The *rijke dijk* (rich dike) is the third initiative: creating tidal pools and a breeding island for birds on the sea side of the dike.

The production site at Bierum, just north of Delfzijl

The Spijksterpompen pumping station that drains into the Eems, with the Eemshaven port in the background

Sample Sheet of Intervention – overview on page 24-27

04

Dike-in-boulevard

Programme: **Second Hoogwaterbeschermings-programma (Flood Protection Programme)**
Province: **South Holland**
Coordinates: **52°06'29.8"N 4°16'23.4"E**
GPS: **52.108263, 4.273169**
Assignment: **coastal reinforcement**
Implementation: **2009–2012**
Water board: **Hoogheemraadschap van Delfland**
Design: **Manuel De Solà-Morales Rubio**
Contractor: **combination Zeefront (Van den Biggelaar and F.L. Liebregts)**
Project Manager: **municipality The Hague and Hoogheemraadschap van Delfland**
Length: **2 km**
Earthworks: **2.6 million m³ sand nourishment**
Other figures: **two diaphragm walls: 85 m x 1.2 m x 20 m; boulevard at 10 m above Amsterdam Ordnance Datum**

Scheveningen is an old fishing village on the North Sea that developed into The Hague's fashionable seaside resort in the nineteenth century. The boulevard is one of the busiest and most attractive places on the Dutch coast. The shock was therefore great when it was shown that this flood defence would be ineffective during a super storm with waves up to 8 metres high. Of course, the chance of such a storm occurring is tiny, 1 in 10,000 per year. But if there is one, the damage to Scheveningen, The Hague and the cities in the Randstad would be catastrophic. A vertical solution such as a high dike or a row of dunes was the most obvious way to protect Scheveningen. But a resort of such repute must have an unimpeded view of the sea. And the Netherlands would not be the Netherlands if it could not find an inventive solution to the conundrum: the dike-in-boulevard was born.

Beach as Breakwater

In 2010 the beach between the Vissershaven and the Scheveningse Pier was made 40 to 70 metres wider: 2.6 million m^3 of sand was spouted onto the beach over a length of 2 kilometres. This measure alone will reduce the height of waves during a super storm from 8 to 1.4 metres. Before the waves even reach the dike they will lose most of their force by hitting the sand first.

Dike Hidden in a Boulevard

The second step involved the construction of a one-kilometre-long dike in the boulevard. This unique technical tour de force is hidden from view in the boulevard itself. The lowest layer consists of geotextile covered with blast furnace slag. Concrete columns were installed on top of this layer, fixated on the top and bottom with liquid asphalt for half a metre. They were mutually stabilized by filling the cavities between the columns with stones

Before: construction of the robust dike that will be concealed under the boulevard, 2010

21

After: The dike is no longer recognizable as such, concealed as it is under the boulevard and the marram grass

33

and then washing them in with sand. At each end of the dike two diaphragm walls, 1.2 metres thick and 85 metres long, were installed in a small deep slot. These reinforced-concrete diaphragm walls were placed more than 20 metres into the ground and provide a connection with the existing sea defence. The boulevard itself has been transformed into a spacious balcony 10 metres above sea level. The gently sloping line of the dike-in-boulevard's design also contributes to the construction's strength: high waves cannot crash against the dike.

Spanish Architect Translates History into Future

The functional variation in height and in distance to the coastline was one of the tasks to be addressed by Manuel De Solà-Morales Rubio. The architect, who died in 2012, had already designed waterfronts and boulevards in cities such as Barcelona and Porto. He proposed restoring the old undulating coastline of Scheveningen by using the curvature of the old dune landscape and the winding line along the coast formed by the buildings' façades. Scheveningen Bad, beach and village are now once again a harmonious unit. The different heights have been used to separate pedestrians, cyclists and motorists as much as possible.

In the old situation the view to the sea was impeded by an almost solid wall of beach pavilions; De Solà-Morales solved this by re-distributing them over five clusters. The spatial quality was also enhanced by the choice of materials. At various points marram grass was planted, giving the new coastal flood defence a dune-like appearance. When the boulevard was opened in 2013 the Minister of Infrastructure and the Environment Melanie Schultz van Haegen quoted the well-known Hague folk song 'Oh oh Den Haag': 'The Hague isn't only a beautiful city behind the dunes but also once again a beautiful and safe city *in front of* the dunes.'

Debris Weighing a Maximum of 700 kg

Building on the dike, in this case on and around the boulevard, is not permitted due to safety considerations. Kiosks, pedestrian bridges, walls, beach pavilions, artworks and the like must therefore be constructed in such a way that they cannot damage the dike during a super storm. If they collapse, they must disintegrate into debris that is too light to cause harm. Drop tests using concrete and steel weights conducted on a dike in Zeeland showed that the maximum permissible weight is 700 kilogrammes.

Periodic Underground Testing

The underground construction is not as easy to inspect as standard dikes. Provision was made for this: a built-in monitoring system with extremely sensitive meters can immediately detect the slightest subsidence of the dike. Various flux towers have also been placed on the dike revetment that record any settlement or subsidence of the dike. And, finally, a small reference dike using the same specifications was built under the beach. It will be dug out every five years to be inspected.

Before: The old promenade along the beach, 2008

After: The new boulevard has been given Spanish flair

Scheveningen: from Fishing Village to Bathing Resort

Scheveningen, a village in The Hague, was built on young dunes that developed in about the twelfth century along the North Sea coast. The counts of Holland settled here during the thirteenth century, and their love of fish soon drew fishermen to the area. Scheveningen has been struck by storm surges several times in its history. During the Allerheiligenvloed (All Saints' Flood) in 1570 half of the village was consumed by the waves. Fishing shaped the appearance of Scheveningen for hundreds of years, witnessed by Hendrik Willem Mesdag's late nineteenth-century paintings. The vast Panorama Mesdag (1881) can still be admired in The Hague: a painting in the round, 120 metres long and 14 metres high, it depicts Scheveningen as seen from the Seinpost dune. From 1800 a bathing resort developed just north of the fishing village, and Scheveningen became a popular resort. Nowadays 17.5 million visitors a year come to the boulevard to sunbathe, relax or visit the sculpture garden at the Beelden aan Zee museum.

This detail from the famous Panorama Mesdag clearly shows the bay in the row of dunes that provides shelter for the fleet of fishing boats

The trajectory of the upgraded boulevard between the Kurhaus and the port with its reintroduced curvature

Passers-by and children in the spring sunshine

The unusual light fixtures partly determine the boulevard's character. Between them, the sculpture of a Scheveningen fisherman's wife, who looks placidly out to sea

The new boulevard is a popular leisure-time location for residents and tourists

Sample Sheet of Intervention – overview on page 24-27

Parking under the dunes

Programme: **Hoogwaterbeschermingsprogramma (Flood Protection System)**
Province: **South Holland**
Coordinates: **52°12'09.1"N 4°23'30.9"E**
GPS: **52.202531, 4.391919**
Assignment: **coastal reinforcement**
Implementation: **2008–2015**
Water board: **Hoogheemraadschap van Rijnland**
Designer(s): **Arcadis (concept and design coastal reinforcement) RHDHV (parking garage and image quality plan), OKRA Landscape architects (public space and entrances parking garage)**
Contractor/consortium: **Bouwcombinatie Ballast Nedam - Rohde Nielsen**
Measurements dune reinforcement: **1,500 m, of which 1,200 m dike-in-dune**
Details parking garage: **500 x 30 m**

Katwijk is a fishing village on the North Sea, situated near the mouth of the Oude Rijn (Old Rhine). The village is about 15 kilometres north of Scheveningen (see p. 94), near Leiden. The history of Katwijk can be traced back to at least the Romans. The relationship between the village and the sea was so intense that until only recently 3,000 Katwijkers lived outside the dike. The sea dike where the old village was located was not on the coast but ran straight through the village. The row of dunes in front of the beach was too low and too narrow to act as a primary flood defence. During the Second World War a 150-metre strip of building along the coast was demolished to make way for the Atlantic Wall. A defensive wall was built on the boulevard. The Andreas church (also called the old or white church) was allowed to remain standing, albeit without its spire. After the war Katwijk was rebuilt as a bathing resort. It was the weakest point along the South Holland coast to be tackled within the framework of the Hoogwaterbeschermingsprogramma (Flood Protection Programme).

Village Brought within the Dike

The existing flood defence was difficult to reinforce, given the village's location, and it was also incapable of protecting the outer-dike buildings. The logical solution was to extend the flood defence to the coastline by reinforcing the row of dunes. Concealing a dike in the dune near the village centre meant it would not impede the view from the beach boulevard. The dunes could therefore remain as high as before: about 8 metres; they were raised to 11 metres outside the village centre.

The old coastline of Katwijk with the distinctive Andreas church, 2009

21
22
26

Parking under the Dune

Reinforcing the dunes was seized upon as a chance to solve the village's parking problem. A 500-metre-long parking garage was built against the built-in dike. This has greatly improved the quality of the public space in the centre and accessibility to the beach. The dunes were temporarily removed during construction. The new dike was built up out of a sand core and covered with stones. The concrete housing for the parking garage was built here on the land side, after which it was completely covered with a dune landscape. The garage has 663 parking spaces, entrances at either end and five exits to the beach for pedestrians. In the new situation, the row of dunes has been considerably widened, to 120 metres. The beach was extended some distance into the sea when sand was spouted onto it from dredging ships (sand nourishment). A shell footpath winds its way through the new dunes. Beach entrances are located near every beach pavilion and in four places in the village centre. The main flight of steps to the beach on the beach boulevard is instantly recognizable because a square was laid out here. A second square is located on the beach side, which can be used for public events.

The Archaeology of Katwijk

Katwijk is an archaeological treasure-house. The mouth of the Oude Rijn in the North Sea was part of Limes, the northern border of the Roman Empire. Katwijk aan Zee was not established until the twelfth century and later developed into an important fishing village. The next economic cycle started in the nineteenth century: tourism. During the Second World War many old buildings were demolished when the Atlantic Wall was built. The construction of the dike-in-dune was a unique opportunity to carry out archaeological research into the rich history of Katwijk aan Zee. Desk research and *in situ* studies were followed by excavations at three locations: at the drainage canal where the Oude Rijn flows into the sea; next to the church; and at the site of the new parking garage. Expectations were high. Brittenburg, the Roman military settlement near the mouth of the Oude Rijn might be found, or a section of the defensive Atlantic Wall. Some even hoped that traces from the Ice Age or other periods of ancient settlement might be discovered. The results, however, were disappointing. Besides small finds dating from the Middle Ages and later periods, the most exciting discovery was, indeed, a remnant of the Atlantic Wall lying upside-down next to the church, camouflage paint and all. The famous Brittenburg has kept its secrets.

The new dike (left) with the housing of the parking garage (right), 2014

The parking garage is enveloped by the new dune, 2014

Interior of the parking garage with easily identifiable symbols

The entrance to the parking garage

The new and wider Katwijk coastline, criss-crossed by cycle and rambling paths

Parking garage entrance and exit as undulating component of the dune

The design for the coastal reinforcement of Katwijk has won several prizes

Cultural History

The Netherlands is a cultural landscape, created out of a centuries-long interaction between humans and the delta they settled in. Step for step the land was ordered and cultivated. The nation's history and the development of the land have both been determined and coloured by the water. Nearly all of our World Heritage sites are related to water: a sea, an island, polder windmills, a steam-driven pumping station, a polder, a canal city, and the water defence line. Water is in our genes, it is intangible cultural heritage. It has formed the Netherlands' culture and determined its identity. The Dutch mentality is rooted in the sea, the polders and living with water. Democracy began with polder authorities, the *hoogheemraadschappen*: we had a *dijkgraaf* (dike reeve) and not a duke. Elections are still held for the bodies that manage our water, some retaining the name *hoogheemraadschap*, others now called *waterschap* (water board).

The Domains of Cultural Heritage

The cultural–historical significance of water and flood defence is expressed in different domains. First, there are of course the tangible monuments, which in one way or another recount the history of life in the delta: windmills, pumping stations, dikes, polders and river aspects. We prefer to preserve and use all this cultural heritage, which does not mean we shy away from transforming it and giving it a new use. The second expression of this cultural–historical heritage is the water system as a whole: the spatial logic and cohesion of the 'water machine' that the Netherlands in fact is. The DNA of the Netherlands can be traced in the structure of the landscape, and so can specific local and regional qualities. This system and structure are important cultural heritage but do not have to be preserved in a literal sense. It is about continuity, building on the existing logic. The ambition is to keep areas recognizable but also provide space for something new. The third domain is the culture of water: the way in which land is made and organized and how citizens are involved in this.

Cultural Heritage Policy

The Cultural Heritage Agency does not aim to 'save' everything old but wishes to contribute to dialogue about the role of cultural history in water projects. Ellen Vreenegoor, a senior adviser on cultural heritage and spatial planning: 'Our hydraulic-engineering cultural heritage goes back centuries and forms the identity of our country, in fact its *raison d'être*. At the same time we notice that expertise about this subject is not automatically included in flood defence and spatial design projects, where speed and procedure are paramount. This sometimes leads to incomprehension among citizens and to expensive planning procedures and appeals. The stratification of the landscape can tell us so much. When we share our expertise people are very enthusiastic. As a nation of water-builders we have, over centuries, made fantastic discoveries that work *with* the landscape instead of against it. Take the phenomenon of the *beemd* [meadow], river banks that can be flooded. Can't they sometimes be returned to their original function?'

Ambitions

The Cultural Heritage Agency as a dialogue partner in water projects aims to contribute information (expertise) and inspiration. At concrete locations the agency stimulates research, preservation and continuity. It is not solely concerned with preserving historic sites but also with grafting new developments onto existing features. At an early stage of planning preparations for large hydraulic works the cultural–historical values are inventoried, ways are sought to integrate special locations and objects into local plans, and spatial quality frameworks for development are formulated. Guidelines on spatial quality, the 'Handreikingen Ruimtelijke kwaliteit', have been formulated for the Rhine, Waal and IJssel rivers, describing what is characteristic of the river in question and which spatial logic determines the river landscape's composition. These guidelines must be taken into account when drawing up flood prevention interventions and spatial designs. In the words of Hein Pieper, *dijkgraaf* (dike reeve) of the Waterschap Rijn en IJssel water board: 'Our work forces us to reach back to the old relationship between landscape and water'.

Yield: Objects and Relics

Recent hydraulic works have provided a large and diverse cultural-historical yield. Archaeological research has led to new insights about the history of civilization in the Netherlands, from the distant past (the prehistory of Limburg) to the Second World War (the Atlantic Wall in Katwijk) and later. In Kampen a perfectly preserved mediaeval cog was discovered in the riverbed and salvaged. Objects have been restored and given new uses, such as a chicory factory on the Vecht river in Dalfsen, farms on the IJssel, and remnants of brick factories along the Rhine. But it was also possible to include ruins of hydraulic works in their new contexts, including the old Hondbossche Zeewering, which, now safely hidden behind the new row of dunes, is a listed historic site. At the outflow structure at Veessen-Wapenveld a *wiel* was left undisturbed, an old round whirlpool created in the distant past by a dike breach. Along the Zeeland coast innumerable elements of cultural-historical significance were included in the new stone revetments, such as the Wachthuisje in Hoedekenskerke and the *Muralt* walls, old concrete dike raisings dating back to the early twentieth century, named after their designer, the baron, engineer and politician Robert de Muralt (see p. 66).

Yield: Structures

Excavating the floodplains and digging secondary channels literally exposes the history of a river. *Reliëfvolgend ontkleien* ('contour-guided clay excavation') uncovers the old sand bed with its contours of channels and ridges. The old coastline of Scheveningen with its characteristic curvature was restored when the boulevard was redesigned. Scheveningen Bad, Strand and Dorp – the bathing resort, beach and village – have been reunited as a cohesive unit. (see p. 94). Cultural heritage has been included in different

The Woudagemaal in Lemmer is a UNESCO World heritage site, and is occasionally used for water management

The cog is raised from the bed of the IJssel river at Kampen

ways in its new flood-protection context – at Fort Ellewoudsdijk in Zeeland, for example, (see p. 66) and Fort Steurgat in the Noordwaard (see p. 140). And, inversely, flood protection was improved through the preservation of a historical context, for example, the specific technical solutions that were found to preserve city and village aspects in Spakenburg, Capelle and Kampen (see p. 206, 196, 200). In the Diefdijk it became possible to experience the structure and function of the Nieuwe Hollandse Waterlinie (see p. 78).

Yield: Stories and Traditions
Intangible cultural heritage is present in different ways in projects. It often concerns forms of participation – the *verhalenjutters* ('story beachcombers') who go foraging for old stories in particular areas, theatre performances, publications, exhibitions and events. In Zeeland an amateur historian is working with the Cultural Heritage Agency to meticulously map the Zeeland dikes, inevitably drawing out stories about the dikes, the landscape and the people. In the redeveloped areas residents and visitors express great curiosity about local stories. Thematic routes, exhibitions, apps and multimedia recounting an area's history and folklore cater for this growing interest about the past.

Remnants of a brickworks in the Ooijpolder along the Waal

Fort Steurgat, part of the Nieuwe Hollandse Waterlinie, was provided with a wave-reducing osier bed (see p. 141)

Controlling the Flow

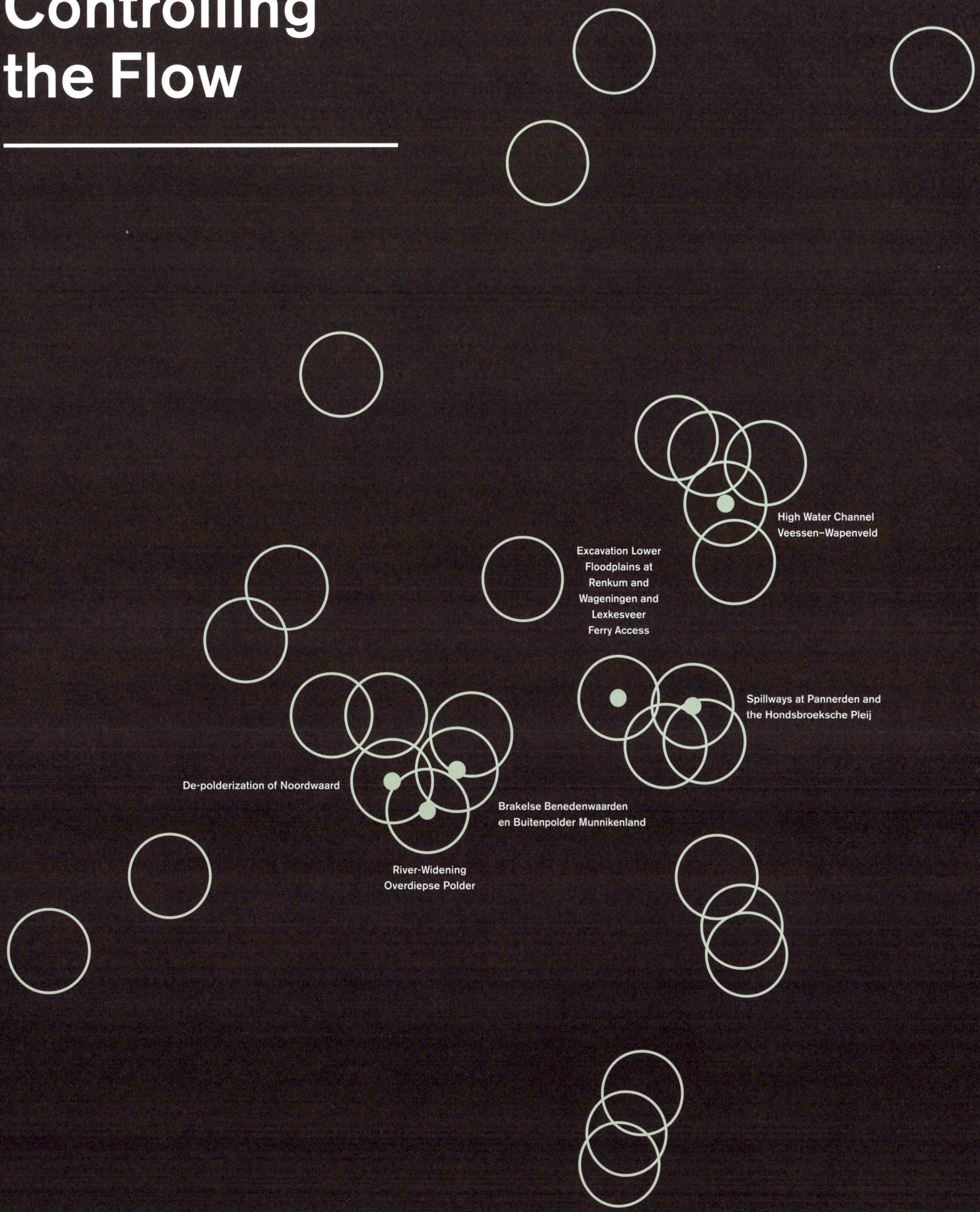

Floodplain excavations Renkumse and Wageninger Benedenwaarden (floodplains) and Lexkesveer ferry access – Inundated floodplains during high water in the Lower Rhine at Randwijk, 2011

High water channel Veessen-Wapenveld – The new dikes for the high water channel (right) are more than 8.5 kilometres long. They are lying quietly in the vast landscape at the moment, but are expected to enclose an immense landscape once in a lifetime, 2016

Depoldering Noordwaard – Viewing point on the roof of a new pumping station, 2016

Sample Sheet of Intervention – overview on page 24-27

Excavation Lower Floodplains at Renkum and Wageningen and Lexkesveer Ferry Access

Nature development as overture to improved through-flow

Programme: **NURG (Nadere Uitwerking Rivierengebied) and Ruimte voor de Rivier (Further detailing river areas)**
Province: **Gelderland**
Coordinates: **51°57'31.5"N 5°41'24.1"E**
GPS: **51.958750, 5.690028**
Assignment: **river widening at the Lexkesveer ferry crossing**
Implementation: **1998–2009**
Water board: **Waterschap Vallei en Veluwe (north bank), Waterschap Rivierenland (south bank)**
Design: **Joris Smits, Royal Haskoning DHV (ferry bridge)**
Contractor: **Ballast Nedam**
Project Manager: **north bank: Staatsbosbeheer; south bank: Gebiedscoöperatie Tussen Rijn en Waal (TRW)**
Drop in water level: **11 cm**
Earthworks: **435,000 m^3: 2/3 for high water shelters (ecology), 1/3 removed**

The Wageningse Berg (Wageningen hill) is only 42 metres high but this most southerly point of the Veluwe lateral moraine provides a spectacular view of the river areas. Two different landscapes meet here: the Veluwe (sand) and the Rivierengebied (clay). Like the river cities, the lateral moraine also creates a bottleneck in the river, partly because of the presence of elevated roads leading to ferry crossing-points in the floodplains. Within the framework of the NURG nature development programme for the river areas, the Lexkesveer ferry crossing was tackled between 2007 and 2009. This project, like other NURG projects at the time, was included in the Key Planning Decision Room for the Rivers. Flood defence therefore also became one of the project's aims and it was expanded to include measures to improve through-flow.

Unhindered Ferry

A combination of measures was implemented on either side of the river: excavating floodplains, digging a secondary channel, creating seepage marshes and removing obstacles. The greatest obstacle was created by the elevated ferry accesses to the Lexkesveer ferry, which literally hindered the through-flow on both banks. During high water the river could only use the summer bed. A river-crossing had existed here even in prehistoric times, witnessed by the ancient routes across the Veluwe that terminate here. The ferry was mentioned in medieval documents and the service is still operating today. As part of the river-widening project, the southern elevated ferry access through the floodplains was levelled and replaced with a bridge that extends over the floodplains to the ferry crossing. The bridge has ten pairs of piers and is 250 metres long – longer than the river's width in its summer bed.

Nature Development

In the floodplains, agriculture partly made way for nature in the shape of seepage marshes and environmentally friendly river banks. The implementation of this aspect of the project took longer than expected because of unforeseen and serious soil contamination that had to be dealt with first, as well as objections and appeals regarding the excavationpermit for the north bank. A bird island was created on the south bank between the river and the secondary channel. *Terpen* (artificial mounds) were built up in the nature reserve to provide shelter for animals during high water. It later appeared that the interventions had been less effective in lowering the water level than had been predicted: only 11 centimetres instead of 18 centimetres. This led to questions in parliament. However, it turned out that the Room for the River target for the required discharge capacity of the Rhine had easily been met thanks to the cumulative effects of the other projects along the river. Moreover, the dikes at the Lexkesveer crossing were high and strong enough, making further interventions unnecessary.

The bridge to the Lexkesveer ferry ramp in summer, 2014

The bridge during high water in winter, 2011

Day-trippers wait for the ferry so they can cross the Lower Rhine

The floodplains of the Lower Rhine create a large, interconnected nature area; the Wageningse Berg (hill) can be seen in the background

The ferry bridge that crosses the secondary channel and the flood-plains across from Wageningen

Sample Sheet of Intervention – overview on page 24-27

Taps for Waal, Rhine and IJssel rivers

Programme: **Ruimte voor de Rivier (Room for the River), NURG (Rijnwaardensche Uiterwaarden)**
Province: **Gelderland**
Coordinates: **51°57'08.4"N 5°57'21.2"E**
GPS: **51.952328, 5.955878**
Assignment: **increase capacity, regulate water distribution among rivers**
Implementation: **delivery Pleij 2010, Pannerden 2014**
Water board: **Waterschap Rijn en IJssel**
Design: **Rijkswaterstaat**
Contractor: **Van den Biggelaar**
Project Manager: **Rijkswaterstaat**

In the border area between Germany and the Netherlands is a relatively low-lying plain between the Montferland, the Veluwe and the Kleef-Nijmegen lateral moraine. This is where the Rhine makes its way through the landscape and splits into the Waal, the Lower Rhine and the IJssel rivers. Over centuries the rivers' course in this area has often changed due to sedimentation and human intervention. Until 1707 the Rhine branched north just past the German city of Emmerich, and it was actually the Waal that entered the Netherlands near Lobith. The construction of the Pannerdensch Kanaal relocated this bifurcation to the Netherlands. The canal, 6 kilometres long and 135 metres wide, supplied water to the Lower Rhine and shortened the navigation route. It also made it possible to regulate the distribution of water between the Waal and the Lower Rhine better. In the nineteenth century Pannerden became an important defence and a fort was built next to the entrance to the canal. The fort was intended to halt advancing troops and, above all, prevent the Pannerdensch Kanaal being dammed off. If that happened, the supply of water to the Nieuwe Hollandse Waterlinie defence line would be cut, making it impossible to defend the west of the Netherlands (see p. 78).

Always One-Third and Two-Thirds

The distribution of water over the three rivers has been precisely fixed since the eighteenth century: at Pannerden the Waal receives twice as much water as the Pannerdensch Kanaal. The canal bifurcates at Westervoort, where twice as much water flows into the Lower Rhine as into the IJssel. It may appear trifling but this

10
14
15

Historic map of the Pannerdensch Kanaal dating from 1701, when it had just been dug

View of the spillway in the excavated Hondsbroeksche Pleij where the IJssel splits from the Rhine (left); Arnhem can be seen in the background

I. Deel. Fol: 138.
De Lijmers.
De Lymersche Bandijk
Candia
De Rivier de Neder Rijn
Peppelgraaf
Het Hanckien
Het Nieuwe gegraaven Retranchement, 1701. off Land
Strang off Laagte daar wel eer de Rivier heeft geloopen.
Vierhondert Rynlandsche Roeden.
10 50 100 200 300 400
Het Bergsche Hooft
De Geldersche Weerd
De Bandijk van Over Betuwe
De Over Betuwe
De Panderfsche Dijk
Doornenburg
Brouwerij
Hulhuizen
Kwaaden Dijk
Panderfsche Dijk
Panderen
Molen
De Rivier de Waal.
Weerd van de Graaf van den Berge
Aanwassende Rijnweerd
Nieuwelijx gegraavene
DOORVAART bij PANDEREN,
Getrokken
Van de WAAL in den RIJN;
Zijnde in de over Betuwe
in Gelderland.

fixed key to water distribution is crucial to protecting millions of people from flooding. If too much water accidently enters one or other of the rivers, it can lead to severe flooding downstream. As part of the Room for the River programme the through-flow capacity was increased at Pannerden and Westervoort through floodplain excavation, the construction of secondary channels, and dike repositioning at Pannerden. There was a danger that these measures might cause too much water to flow into one river in particular. Projects equalizing the high water levels of the Waal, Lower Rhine or IJssel downstream might also disrupt the distribution of water between the rivers. In short, a sort of 'tap' was urgently needed in the floodplains at Pannerden and Westervoort to regulate the flow of water between the different rivers when water levels fluctuate. Hence the 'flexible spillways', which were among the first projects to be implemented as part of Room for the River. In Pannerden the new spillway replaced the existing spillway; in the Hondsbroeksche Pleij a spillway was built and the dike was repositioned.

Concrete Components

The flexible spillway at Pannerden is a 175-metre-long concrete structure in the floodplains, perpendicular to the river. The structure consists of 32 gates, 5 metres wide and 5 metres high. On top is a sort of pergola whose concrete beams (stop logs) can be lowered with a crane to close off the gates. Each beam is 5 metres long and 1 metre high. The flood defence is innovative in its simplicity, without automatic valves, motors or control mechanisms. This low-tech solution requires little maintenance and is unlikely to break down. It resembles the flashboards that have been used in dike drainage canals for centuries. The size of the spillway's through-flow openings is re-calculated every year to ensure an optimal water distribution. A crane can be used to lift beams in and out of the spillway. This makes it possible to meticulously regulate the amount of water flowing through the spillway into the Pannerdensch Kanaal. The spillway in the Hondsbroeksche Pleij at Westervoort regulates the flow of water from the high water channel to the IJssel in the same way. This spillway is 150 metres long and includes 30 gates. If necessary, the through-flow can also be adjusted here during high water using a crane transported on a pontoon.

Solution to Seepage

The spillway in the Hondsbroeksche Pleij was completed in 2012. Measures were also taken to deal with a second problematic body of water: seepage. A common problem during high water is that the increasing pressure created by the river water causes groundwater to rise to the surface. A waterbed effect causes seepage to flood cellars and creeping spaces far behind the dikes. When the dike was repositioned, a 9-metre-deep seepage cut-off wall was installed in the new dike body to prevent the dike from leaking. For the first time, the wall was made of bentonite, a fine clay that waterproofs the soil. To prevent seepage causing problems in nearby Westervoort, the little old Stang river was turned into a 'seepage window': the river was deepened and widened, turning it into a water circuit that stores the seepage water and reduces pressure on the ground-water. A small pumping station empties the seepage water into the IJssel, thus solving the water problem. Constructing environmentally friendly banks for the Stang also improved its aspect and natural values.

The spillway in the Hondsbroeksche Pleij shortly after completion, 2010

Before: The old situation in the floodplains next to the canal at Pannerden, 2010

The spillway through-flow opening can be adjusted with these stop logs

After: The spillway improves water-flow control in the 'green river' when the water level is high

The spillway at the Hondsbroeksche Pleij, high water 2012

Sample Sheet of Intervention – overview on page 24-27

A mega intervention with minimal impact

The IJssel river is perhaps at its most beautiful between Deventer and Zwolle, with winding dikes, old villages, panoramic views, the silhouettes of flourmills towering above the flood-plains and, in the distance, the Veluwe lateral moraine. This idyllic landscape is the setting for a tour-de-force high water channel. When this bypass fills up, the water level in the IJssel will (temporarily) drop by 71 centimetres, and the effects will be noticeable as far upstream as Zutphen, 35 kilometres away. This 'spare river' cuts off an enormous bend in the IJssel and is located outside the river channel. Fertile inner-dike farmland from the Veesser-, Vorchter- and Wapenveldse-broek had to be annexed to realize the project. This probably explains the intense opposition to the project at the time. Considerable time and energy were invested in a collective consultation process with residents, companies and other interested parties. This led to increased attention to spatial quality and integral regional development in which flood protection was linked to agriculture and recreation. Many of the parties now not only endure the inevitable disruption caused by the intervention but also benefit from it.

River on Dry Land

Unlike most river projects, spatial planning for the high water channel did not include

Programme: **Ruimte voor de Rivier (Room for the River)**
Province: **Gelderland**
Coordinates: **52°22'23.9"N 6°04'06.5"E**
GPS: **52.373302, 6.068470**
Assignment: **reduce flood risk along the IJssel river during extreme high water**
Implementation: **project decision March 2012; commissioning February 2017**
Governments involved: **municipalities of Heerde and Olst-Wijhe, Province of Gelderland**
Water board: **Waterschap Vallei en Veluwe**
Designer(s): **Veenenbosch en Bosch landschapsarchitecten (target plan spatial quality), Lodewijk Baljon landschapsarchitecten (landscape design), ZUS (design structures), Iv-infra (water control works, valves)**
Contractor/consortium: **Combinatie IJsselweide: Boskalis, Van Hattum & Blankevoort.**
Stakeholders: **residents' advisory group, Natuurbelang, Kamer van Koophandel/KvK, Plaatselijk Belang Marle, Veluwe IJsselzoom, Vereniging Dorpsbelangen, Heerder Historische vereniging, Platform Recreatie en Toerisme.**
Drop in water level: **71 cm**
Surface project area: **surface channel approx. 720 ha**
Earthworks: **Construction 16-km dike: approx. 1,200,000 m³ sand and 500,000 m³ clay. Approx. 25 km drainage canals: approx. 500,000m³ earthworks, bridges, pumping stations, culverts: 11,000m³ concrete and 1,500 tonnes reinforcement**

Each of the distinctive concrete portals used for the inlet structure was cast separately, 2016

a transition from agriculture to nature development; the existing land-use situation remained largely unchanged. The impact of this enormous intervention was therefore kept to a minimum, both during construction and, following completion, on spatial use and activities. The high-water channel is intended for emergency use only. In normal winters the inflow structure remains closed and no water flows into the channel. The flaps will only open here if the level of the IJssel reaches 5.65 metres above Amsterdam Ordnance Datum. The chance that this will happen is 1 in 80 per year, that is, once in a human lifetime. The area had to be drastically overhauled in order to add two 8-kilometre-long dikes. The channel is between 500 and 1,500 metres wide and was not dug below ground level but, rather, constructed on ground level. Roads remained in place, and normal agricultural work has continued uninterrupted in this high-water bed. Only the trees, shrubs, buildings, cables and lines were removed, and nine farms and five houses were demolished. Transmission pylons were lifted up and placed on metal plinths so that they do not hinder the through-flow or become damaged. When the high water channel is used, an island with about 1,000 inhabitants will be created between the channel and the IJssel. It will remain accessible via bridges that have been built at each end of the high water channel.

Inlet Structure

The inlet structure was built under the Tolbrug in Kerkdijk in Veessen. It consists of 60 steel locks over a total length of 785 metres. Under the bridge a fascinating space was created by the long row of asymmetric concrete portals that gently curve away in line with the bridge – massive next to the locks and slender on the high water channel side. Before the high water channel is used, bright red yokes and cylinders (nicknamed *melk-meisjes*, or 'milkmaids') will be hung from the bridge piers. Using a mobile battery, the locks can be pulled up one by one like the visor of a helmet. Each lock is 12 metres long and 1 metre high. For a flood defence that is expected to be used once in a human lifetime (80 years) it is crucial that everything works at the critical moment. The confidence placed in simple mechanisms rather than complex technology is striking: they are reliable and easy to maintain. Four men can activate the channel within 18 hours at the most. The water will rush forcefully into the high water channel. Concrete stilling basins behind the first 13 locks that open can store the water and guide it into the area. After two hours there will be enough water in the channel and the other 47 locks can also be opened. The bed of the whole inflow zone has been reinforced and covered with soil and grass to prevent damage and erosion. After use it will only be necessary to add new soil and grass. A temporary loading and unloading quay was located in the IJssel near the village of Marle during the construction of the dikes, bridges and other structures. Temporary roads were built throughout the channel area so that construction traffic caused hardly any disruption on existing roads. The sand was delivered from the unloading quay using a pressure pipeline.

The 17 high-voltage transmission towers in the high water channel are placed on an elevated foundation to protect them from high water in the channel, 2013

The Tolbrug inlet structure stretches over the entire width of the high water channel. Right, the village of Veessen on the IJssel

Outfall Structure

The water will flow towards Wapenveld via the high water channel between the new dikes, which are between 3 and 5 metres high. Next to the discharge structure are two bridges that are separated by a *wiel*, a small round lake directly behind the old dike formed when water rushed in during a long-ago dike breach. The water can flow into the IJssel again over a threshold that is 4.10 metres above Amsterdam Ordnance Datum. A discharge structure can also be opened to drain the channel and, lastly, a pumping station can normalize the water level again. The locks of the inlet and the threshold at the outfall structure prevent the high water channel filling up unnecessarily. The water will reach the locks during high water about twice a decade, without flowing into the channel.

Regional Development

A landscape zone was created outside the channel to compensate for the lost natural value, and an area for water fowl inside the channel was enlarged by 15 hectares. Land reconsolidation facilitated increases in scale in agriculture. This land re-consolidation, conducted by Stichting Effectief Verkavelen (STERK), was accomplished in less than one year. It enabled several commercial farms to expand and modernize their operations, from the high-tech robotized dairy farm to the ecological nature farm. Determining new uses for the vacant farms in the area is a challenge – options include health-care and recreational initiatives. Regional development has created tourist attractions here, including Erve IJzerman in Wapenveld, where tourists can experience the farm life of a bygone era. Other recreational initiatives include the construction of cycling and rambling paths, a canoeing route, a visitors centre, and shops selling regional and design products. To mark the opening of the high water channel in 2017, residents gave a theatre performance depicting the project from their point of view.

The gates are open, the red lifting cylinders, nicnamed the melk-meisjes ('milkmaids') are clearly visible

The Kromme Kolk (left), the remnant of an earlier dike breach, has been preserved between the high water channel's two drainage openings

The rear of the inlet structure with the reinforced bed to limit erosion

Sample Sheet of Intervention – overview on page 24-27

A successful citizens' initiative

Programme: **Ruimte voor de Rivier (Room for the River)**
Province: **North Brabant**
Coordinates: **51°42'54.4"N 4°56'47.2"E**
GPS: **51.715111, 4.946444**
Assignment: **remove bottleneck in the Bergsche Maas (Bergsche Meuse**
Implementation (from comprehensive planning to implementation): **2004–2017**
Water board: **Waterschap Brabantse Delta**
Designer(s): **Spatial design (terpen plan): Bosch Slabbers in collaboration with Witteveen + Bos, Tauw, Equator and Bugelhajema (embedding plan). Design terp farms: Onix Architecten.**
Contractor/consortium: **Aannemerscombinatie De Hollandsche Waard: Van Oord Nederland BV, GMB Infra BV, Oldenkamp BV Arcadis**
Stakeholders: **Province of North Brabant, Belangenvereniging Overdiepse Polder, municipalities of Geertruidenberg and Waalwijk, Bezinningsgroep Water, Habiforum**
Drop in water level: **27cm**
Surface project area: **approx. 550 ha and 180 ha floodplain**
Earthworks: **excavating, transporting and processing 1,850,000 m³ sand, 1,110,000 m³ clay, 400,000 m³ soil**

Directly to the south of the major rivers is the Overdiepse Polder, bounded by the Bergsche Maas (Bergsche Meuse) and the Oude Maasje (Old Meuse), some 25 kilometres from 's-Hertogenbosch. The area had always been marshy and was flooded every winter. In the 1940s a guide lock was constructed that drained the polder. From 1975 the polder was home to 18 farms. Less than 25 years later it appeared that the Bergsche Maas did not have enough room to cope with predicted water levels. A plan was drawn up to let the polder flood during high water in order to keep upstream cities and areas dry. It entaïled the construction of two secondary channels and an end to farming in the area. It is hardly surprising that residents and entrepreneurs were up in arms. From a tense beginning, a consensus and constructive collaboration developed over many years. Residents drew up a counterproposal that combined agriculture and high water storage. This was the plan that was finally adopted.

10

The Return of the Terp

The residents and entrepreneurs in the Overdiepse Polder fully understood the threat posed by the water. At the same time, they themselves were struggling with far-reaching changes and increases in scale in agriculture. Several residents invited provincial administrator Jan Boelhouwer for consultations. He challenged them to come up with ideas about and requirements for water storage in the polder. The farmers wanted

The Bergsche Maas gradually overflows its banks through the inflow gaps in the summer dike, 2014

19

Regular meetings between all the parties in the contractor's site hut, 2012

to retain and at the same time restructure farming in the area. Two residents, Nol Hooijmaaijers and Sjaak Broekmans, suggested placing the farms on *terpen* (artificial mounds), thus restoring an ancient northern Dutch tradition in the south of the country. The idea was developed further in conjunction with the Zuidelijke Land- en Tuinbouw Organisatie (ZLTO), a regional organization acting in the interests of farmers and horticulturalists. A proposal for a series of approximately eight *terpen* along the south edge of the polder was drawn up. Half of the farmers would probably leave the polder and start a new life elsewhere. An anonymous questionnaire showed that six families wanted to remain, six wished to leave and six were unsure.

Policy Experiment Creates Room for Participatory Planning Process

The farmers had considerable difficulty getting a hearing for their plan. Gradually, the tide turned, in part due to the involvement of administrators, politicians and officials. The *terpen* plan was given extra impetus when it scored higher than the original river-widening plan in terms of both cost and future value. The Overdiepse Polder was designated as a policy experiment, which created room within the usual procedures to work towards alternative solutions in a participatory process. Many of the insights acquired here were later included in the Room for the River projects.

Agriculture and High Water Storage

Following 15 years of participation and deliberation, the Overdiepse Polder was transformed in line with the residents' plans. Everyone took part voluntarily; no land was expropriated. All old farms were demolished. The Bergsche Maas dike was lowered slightly, turning it into a 6-kilometre-long threshold that will be flooded during very high water. The chance of this occurring is 1 in 25 per year. The 27-centimetre drop in water levels helps reduce the risk of flooding in an area inhabited by four million people. After an inundation the water can flow away through an outlet to the Bergsche Maas and a pumping station. The farmers will be compensated after each flood. A high new dike was built along the southern edge of the Overdiepse Polder, with eight farm *terpen* on its outer side. The polder also hosts ecological corridors. When sand-dredging operations ended, the extraction site was partly filled in and a small lake developed that is now part of a 5.5-hectare nature reserve.

Remainers and Leavers

It was difficult to separate public and private interests honestly to ensure that farmers were not advantaged or disadvantaged. In principle, the wealth and income position of all farmers before and after the purchase of their business was to remain unchanged and they were to be compensated for the costs of restructuring. It was also vital to ensure that the remainers would not have to incur large debts in order to extend and modernize their farms. Ten farmers left the polder. Most started new farms in the area, though one family emigrated to Canada. Both the leavers and the remainers greatly improved the quality of their farming operations and are well prepared for the future. The farmland in the Overdiepse Polder was bought by the eight farmers who stayed. As a result of the residents' input, the Overdiepse Polder was not turned into a nature reserve but, rather, a high-quality agricultural area. The participatory process lasted 15 years and caused many headaches but it was ultimately decisive for the project's success.

Ribbon of farms

Bosch Slabbers drew up a landscape design for the south dike with its eight *terpen*. The *terpen* consist of a lower *terp* and an upper *terp*. The lower *terp* ensures that the farms are self-contained in relation to the dike. They will be slightly separated from it by a ditch. A double row of ash trees line three sides of the lower *terp*, with tree-filled meadows or orchards on the dike side. The houses have been built on the upper *terp*, with stables and barns at the back. A competition was held for the design of the buildings, which was won by Onix Architecten. The assignment was to come up with a twenty-first-century ribbon of farms. The aim was to create an ensemble, a 'farm family'. Emphasis was placed on how buildings should be organized on the farm and on the choice of façades, materials, roof-shapes for the houses and the like. The farmers were also free to develop their own parallel plans. The family connections between the farms gave the dike its unique character, and it is now a tourist attraction. A new trend has even developed in Brabant: *terp* tourism.

Milking time: the cows walk from the pastures to the terp and head for the cowshed

The Overdiepse Polder after completion of the project. The new terps are neatly lined up on the left along the dike

Construction of a sturdy new terp, 2011

Sample Sheet of Intervention – overview on page 24-27

10
14
15
32

Brakelse Benedenwaarden and Buitenpolder Munnikenland

A grandstand in the landscape

Programme: **Ruimte voor de Rivier (Room for the River)**
Province: **Gelderland**
Coordinates: **51°48'43.6"N 5°02'23.0"E**
GPS: **51.812111, 5.039722**
Assignment: **improving through-flow during high water**
Implementation: **2007–2016**
Water board: **Waterschap Rivierenland**
Designer(s): **Royal Haskoning / DHV BV, Kees van der Velden landschapsarchitectuur, Bureau Waardenburg.**
Contractor/consortium: **combination Van Oord–GMB (main contractor), Martens en Van Oord (earthworks)**
Stakeholders: **Staatsbosbeheer in collaboration with Free Nature and De Capreton (Vereniging voor Agrarisch Natuur- en Landschapsbeheer Bommelerwaard), Province of Gelderland, municipality of Zaltbommel, Rijkswaterstaat, Stichting Slot Loevestein, Dunea Duin en Water, Vitens, Hengelsportvereniging De Rietvoorn, Regionaal Bureau voor Toerisme Rivierenland, Stichting Ark, Stichting Behoud Waterlinie Bommelerwaard, Stichting de Vier Heerlijkheden, KPN, Liander BV, Ziggo, Wildbeheereenheid Bommelerwaard, Historische Kring Bommelerwaard.**
Drop in water level: **11.5 cm**
Surface project area: **700 ha**
Earthworks: **approx. 2.4 million m^3 excavated**
Real estate acquisitions: **2 farms and approx. 199 ha land**

The Netherlands is a delta made up of three European rivers. The Schelde (Scheldt) flows into two sea arms in Zeeland, in the south-west of the country. The Rhine and the Meuse criss-cross the country in a maze of rivers that branch out and then flow together again. The watercourses have been changed and repositioned constantly over the centuries. Every little stream has its own name. Long ago the central river area formed a hard border in the Netherlands between the Protestant north and the Catholic south. One of the country's most famous castles, Slot Loevestein, is located in the middle of the delta. The Waal and the (Afgedamde) Maas (dammed-off Meuse) merge here and flow on via the Biesbosch swamp to the Hollands Diep and the Haringvliet before reaching the sea. During high water the castle is cut off on a small island, surrounded by a mass of water.

Outer-Dike Nature Area

The outer-dike area was enlarged to increase the Waal's discharge capacity during high water. The new primary flood defence was constructed one kilometre inland to this end. Two secondary channels were dug in the floodplains to collect excess river water during high water. This drastic intervention led to the virtual disappearance of farming from the floodplains, which are now mainly

The terraces of the grandstand dike are taking shape, 2016

The crest of the grandstand dike provides a good view

Immense excavation works in the western floodplain along the Merwede, 2014

The Wakkere Dijk around the excavated Buitenpolder Munnikenland is so wide that it is barely recognizable as a dike from the air

zoned for nature and recreation. The intervention resulted in an 11.5-centimetre drop in water levels. In the Gandelwaard (near Slot Loevestein) clay-dredging by a brick manufacturer is carving out a channel that, when completed, will reduce the water level further.

Multifunctional Grandstand Dike

Most of the soil from the old Waaldijk (Deltadijk) and the floodplain excavation was not removed. Instead, it was reused in the new 2-kilometre-long dike. This dike is between 150 and 300 metres wide, making it considerably wider than most dikes. This over-dimensioning made it possible to use the dike for more than only flood protection. It is the first multifunctional river dike in the Netherlands, comparable with the climate dike next to the Lek (see p. 74)). Terraces have been built on its outer side to create a grandstand. It will never be as crowded as the Spiegelwaal in Lent (see p. 236). The landscape creates a spectacle that the passer-by can watch from the grandstand. When the water is high the dike provides shelter for the cows, horses and other animals that are used to control the floodplain vegetation. Giving the dike a broad inner berm on its land side created room for tree-filled meadows. Cycle and rambling paths were built over the dike body. The focus group (residents and stakeholders) proposed calling the multifunctional grandstand dike the Wakkere Dijk ('the awake dike'), a variant on the names given to the usual string of river dikes: *waker* ('guardian' or summer/primary dike); *slaper* ('sleeper' or winter/secondary dike); and *dromer* ('dreamer' or emergency/tertiary dike).

Cultural History

Slot Loevestein is a major tourist attraction in the area. The Contrescarpgracht moat around the castle was dug anew. This is the third and outer-most moat, outside the castle moat and the fortification moat. The carpark next to the castle was redesigned and the access road was made higher. The site can still be reached during high water. Other traces of the past can be found in the surroundings including forts and bunkers that are part of the Nieuwe Hollandse Waterlinie (see p. 78). The vanished Kasteel Munnikenland, the Kloosterboerderij Munnikenhof and the Rechthuis can now be identified because the sites have been elevated and marked out with a planting scheme. The nature values in the region, which is subject to tidal influences, have been reinforced. Many values are designated Natura 2000 areas because of their ecologically valuable meadowlands.

Around the Slot Loevesteijn the outer moat, the Contrescarpgracht, has been re-excavated

Cyclists zig-zag down
the grandstand dike

Sample Sheet of Intervention – overview on page 24-27

An advanced water machine

Programme: **Ruimte voor de Rivier (Room for the River)**
Province: **North Brabant**
Coordinates: **51°48'00.2"N 4°51'55.3"E**
GPS: **51.800041, 4.865367**
Assignment: **remove bottleneck in the Merwede river near Gorinchem**
Implementation: **2010–2015**
Water board: **Waterschap Rivierenland**
Design: **Volker InfraDesign; design vision: Robbert de Koning**
Design bridges: **West 8, in collaboration with ipv Delft**
Contractor: **Bouwcombinatie Noordwaard: Boskalis, Van Hattum en Blankevoort (design and construction structures), Martens en Van Oord groep (earthworks, demolition and drainage), Gebr. Van Kessel Wegenbouw**
Project Manager: **Gebr. Van Kessel and FREE Nature, commissioned by Rijkswaterstaat**
Drop in water level: **60 cm at Werkendam and 30 cm at Gorinchem**
Surface: **4,450 ha**
Earthworks: **4,000,000 m^3**
Other figures: **70 km new embankments and dikes, 45 km creeks, 29 terpen and bespoke solutions, 38 bridges, 12 pumping stations, 17 windmills, 19 engineering constructions (e.g. culverts), 50 km sealed roads, 4 carparks, and demolition 50 houses and barns**

The Merwede river near Gorinchem is one of the busiest navigation routes in the Netherlands and it is also a bottleneck in the river drainage system. Upstream the Meuse and Waal meet at Slot Loevestein. Downstream the river bifurcates at the Eiland van Dordrecht. The danger presented by the bottleneck at Gorinchem became clear in 1995 when the eastern part of the city had to be evacuated because of the risk of flooding. The Room for the River programme made it clear that the pressure on Gorinchem had to be relieved. But there was no place to go: Gorinchem is situated directly on the Merwede, and the ancient fortified city of Woudrichem lies opposite. However, it became clear that excavating the floodplain at Avelingen would achieve a 12.5-centimetre drop in water levels. This, and creating a high water drainage system 8 kilometres downstream, straight through the Noordwaard polder, would ease the pressure on Gorinchem sufficiently.

Swamp and Tidal Area

The Noordwaard embodies the Netherlands' history with water. It was once dry land but after the famous Sint Elizabethvloed deluge in 1421 it became an inland sea with swamps, of which the Biesbosch is a remnant. The land slowly re-emerged as a result of siltation and poldering. A patchwork of small polders developed, carved up by tidal gullies that filled up during high tide. As part of the Delta Works, the Haringvliet was closed off in 1970, limiting tidal movement to 30 centimetres. The whole area was then consolidated into a single large polder: the Noordwaard polder has only existed since 1980.

16
19
28

Living high and dry on a terp during high water in the Noordwaard

Part-Time River

Less than a generation later the water was back, this time as a part-time river included in the Room for the River programme. The inundations are conducted by man, right down to the smallest details. A river bed was created on the floor of the polder, complete with dikes. Even the old creek pattern was partly reinstated, in a slightly enlarged version. The existing river dike along the Merwede was lowered over a 2-kilometre length to 2 metres above Amsterdam Ordnance Datum. This dike was covered with clay and grass and functions as a threshold. As soon as the water exceeds this level, the dry riverbed comes alive: the water flows through the Noordwaard via the Biesbosch to the Hollands Diep. The dikes in the polder, which are referred to as *kades* (embankments), are not all the same height. Unlike the Veesen-Wapenveld high water channel (see p. 126), the low-embankments polders in the Noordwaard fill up with water every year and nearly half of the Noordwaard turns into a water storage area. As soon as the water level drops, the water flows out of the polders via gate culverts. Windmills and 13 small pumping stations drain the polder.
On the very few occasions when the water reaches extremely high levels, the polders with high and very high embankments will also fill up with water. According to the calculations, the chance of this happening is 1 in 100 and 1 in 1,000 per year respectively. In those exceptional cases, extra (emergency) pumps will be needed to drain the water afterwards.

Water Machine

The Noordwaard, with all its high and low embankments, thresholds, gate culverts, windmills and pumping stations has become an advanced water machine. Little human intervention is required; the water finds its own way. A system consisting of 38 bridges and roads opens up the polder. The routes over the low embankments are mainly intended for agriculture and cyclists. The main traffic connections traverse the high embankments, which are rarely flooded. These embankments provide an escape route out of the polder for humans and animals. The periodic evacuation drills prove that in the Noordwaard people really do work with the water.

Wave-Breaking Stand of Willows

The interventions that have been carried out in the Noordwaard show numerous remarkable solutions that suit the region's unique characteristics. A new river dike was needed very near Fort Steurgat, for example. The fort is inhabited and is part of the Nieuwe Hollandse Waterlinie (see p. 78). To retain the fort residents' unimpeded views and leave its surroundings undisturbed, the dike tapers away here and its crown is 70 centimetres lower

The new creeks help shape the water machine, 2014

A delta in miniature: land and water alternate during high water in the Noordwaard

The water flows further into the area via culverts

Water buffalo as well as Konik horses and Galloway cattle are deployed to keep the vegetation under control, promoting the flow and storage of water

than planned. This could only be done if flood protection could be guaranteed in some other way. The solution was to plant a stand of willows in front of the dike that acts as a wave-break and relieves pressure on the dike. An environmentally friendly wave-breaking dike had never been used as a primary flood defence in the Netherlands before.

Bricks Reused

Farms were demolished in the areas that are flooded every year and the bricks reused for the pumping station walls. A viewing platform was added to the roofs of two pumping stations. New farms in the low-embankment polders were built on *terpen*, the artificial mounds that had traditionally been a flood-protection measure in the north of the Netherlands and that had also been constructed in the neighbouring Overdiepse Polder (see p. 132). In the Noordwaard 12 of the 25 farms were left following the intervention. Some 400 hectares of new nature was developed here. The Netherlands has hardly any fresh-water tidal areas left, making the natural value of this area exceptionally high. Water buffaloes can be observed here – being large grazers, they keep the vegetation under control. The area's tidal system and nature development made a link with the Natuurpark De Biesbosch obvious, witnessed by the modernized Biesboschmuseum on the Museumeiland in the Noordwaard.

Waiting for the Water

Thanks to Room for the River, the improved flood protection for Gorinchem led to a drastic overhaul of the magnificent landscape a few kilometres away. This water-filled landscape gathers together agriculture, nature, architecture and recreation. Floods are no longer a threat but a spectacle that is part of everyday life, one that can be experienced.

Bricks from demolished houses have been reused for the walls of a new pumping station that also serves as a viewing point

The difference in level is clearly visible when one polder is flooded and the other is (still) dry

Participation

Large spatial projects have a huge impact on the places where they are implemented. They can completely disrupt the lives of local residents. People have to move house, refurbish their homes or leave their jobs, or they find themselves living in completely transformed surroundings. But this can also create opportunities: a better home, more room for entrepreneurial activities, new business models and a higher quality living environment. These days in the Netherlands, it is unthinkable that you would ignore the interests of local residents, but water safety remains the top priority. There is always the risk of a stalemate between the public authorities on the one hand and residents and businesses on the other. However, in practice an acceptable solution can nearly always be found if the parties consult with one another. For the people involved, the downside of the nuisance and transformations can be combined with the upside of new opportunities, developments or quality improvements. Participation has progressed beyond the stage of giving lectures, publishing joint newspapers or putting on therapeutic plays. Residents have evolved from victims who need support to serious partners in a dialogue who are prepared to shoulder some of the responsibility.

Human Engineering

In our modern-day society with its unprecedented affluence, it seems only natural that the quality of the surrounding area and the interests of individuals should be taken into account. Inge Lardinois, head of the *Programmering Waterbeleid* ('Water Policy Programming') department at the Ministry of Infrastructure and the Environment: 'Both aspects are so crucial. That's why it so good that the Directorate-General for Public Works and Water Management has a technical manager, a contract manager and an environmental manager. Because sometimes you are forced out, off your land, and that's a severe blow. I attended the delivery of Noordwaard and got talking to a farmer's wife, the first who had left. The family had moved to Zeeland but the children were unable to settle there. I saw the regret in her face, the impact of this move on her life (see p. 14).' Participation is the latest offshoot of the 'inclusive' approach to landscape design and technology.

The Design Bridges the Contradictions

The then National Landscape Advisor Eric Luiten in 2014: 'Agreement on dike reinforcements can only be reached with a persuasive, logical spatial design in which the technology, landscape and community come together in an area-specific and affordable solution.' That leads to meetings in which public administrators compare the various measures for their municipality (dikes, options for reductions in water levels and the costs). Such meetings help spell out the water safety task, explore the feasibility of various solution areas (finding the right balance between river widening and dike reinforcement) and identify opportunities for fitting measures in and combining them with other tasks.
The design has become much less of an engineering solution and more the outcome produced by creatively weighing up the interests of safety, public support and spatial quality. Communication and the process setup have become essential elements of the design.

Opponents Have Become Initiators

The Overdiepse Polder (see p. 132) is an excellent example of participation in a river widening project. Residents and businesses were not happy with the plan to change the designated use of the polder from agriculture to nature and a water storage area. They joined forces, came up with an alternative plan, were able to use a national government grant to flesh it out, and eventually saw their idea put into practice: an agricultural polder with an annual risk of flooding of 1 in 25, and farms on raised mounds. At the same time, the farmland was cleaned up, a number of farmers could move their farms elsewhere and those who stayed were able to substantially increase the amount of land they farmed.

Co-ownership

The Overdiepse Polder illustrates the importance of making the affected residents and businesses co-owners of the transformation. When they are put in this position, they come up with ideas for the transformation, opportunities for combining interventions and compensating measures. That works well in many projects because it seems water safety makes a good combination with other objectives, ranging from nature development to agriculture and recreation. The challenge is to find a balance for each individual resident between the pros and the cons of an intervention. That starts with an open dialogue from an early stage with residents. But more is needed, namely shared problem ownership. For Waterdunen in Zeeuws-Vlaanderen, an entrepreneur in the recreation sector joined forces with Zeeuws Landschap to come up with an idea for a new dune and nature area (see p. 84). In the Harlingen projects, the Hondsbossche dunes and the Lekdijk in Capelle, residents could provide input and they were closely involved in the design (see p. 188, 220, 196).

Co-creation

The challenge of ensuring participation was tackled thoroughly in the dike improvement projects in Gorinchem-Waardenburg and Waarburg-Tiel/Varik Heesselt. It started with a crash course in dike strengthening for local residents. This turned them into a valued partner in talks and gave them the opportunity to present their wishes to the municipal administrators. A recurring theme was the desire to have contact with the river rather than be facing away from it. This paved the way for the Climate Dike, which is wider than is strictly required for safety reasons, and thus future-proof. The dike will be surmounted by a broad boulevard that will bring the village and river closer together. In another dike project, the first step was to produce a book, *De dijk is van ons allemaal* ('the dike belongs to us all'). This involved the Rivierenland water board interviewing 1,100 residents in eight villages about their relationship with the dike and their opinion of the plans. The book shows how local residents feel about the dike. They were also given the opportunity to suggest types of solutions for the water safety assignment. This led to 18 solution areas; three of these alternatives, each with two variants, are deemed promising. An Environmental Impact Report will now be compiled for these options in which the impact of each proposal on the

Residents contribute their views on plans for the Noordwaard, 2006

Consultation evening about plans for a dike reinforcement on the Eiland van Dordrecht, 2009

surrounding area and the environment will be investigated. This will enable an evidence-based decision to be taken.

Physical Participation

Participation is not just about helping think up solutions, opportunities for combinations and designs – it can also have a very physical dimension. Just as many villages have a volunteer fire brigade, so the town of Kampen has its own Flood Brigade with 200 volunteers (see p. 200). As they literally build up the water defence system by working together under the command of the Drents Overijsselse Delta water board, levels of public awareness of flood protection in this IJssel town are probably higher than anywhere else. In the annual practice run in September, at the start of the storm season, the volunteers go through all the steps in their manuals and erect the entire defence system. You would almost think this performance on the waterfront is part of the annual Dutch Heritage Day (*Open Monumentendag*), which is on around the same date. The Flood Brigade is part of the intangible cultural heritage of the Dutch water protection system.

A farm family install the first pile for their new terp farm in the Overdiepse Polder, 2011

Workshop where local residents are consulted on various variants for dike reinforcement at Eiland van Dordrecht, 2009

Earning Money with Gravel

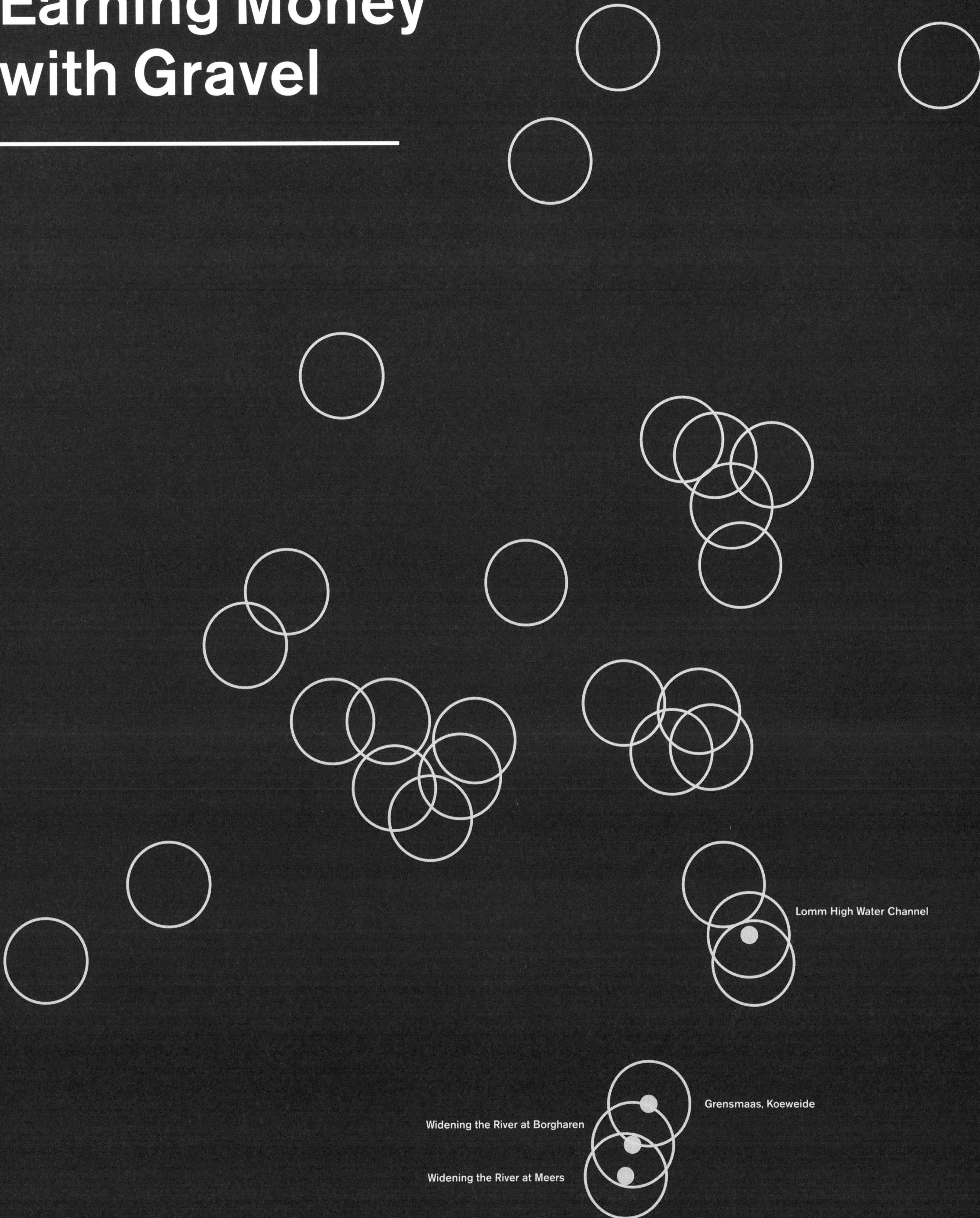

River widening Meers – Sand and gravel extraction site along the Grensmaas, 2010

Grensmaas, Koeweide – Large-scale gravel extraction for vast water storage facility, 2017

The Lomm high water channel – Also created as a result of sand and gravel extraction, 2017

03
Sample Sheet of Intervention – overview on page 24-27
11
14
15

The River Meuse – tamed but still at its wildest

Programme: **Grensmaas Meuse Works**
Province: **Limburg**
Coordinates **50°58'00.2"N 5°44'12.1"E**
GPS: **50.966712, 5.736703**
Assignment: **reduce the high water level**
Implementation: **2000-2025**
Water board: **Limburg Water Board**
Design: **Directorate-General for Public Works and Water Management in collaboration with the provincial authority of Limburg**
Contractor: **Grensmaas Consortium: contractors united in Kombinatie Regenboog (Van den Biggelaar Aannemingsbedrijf BV, Van Oord Nederland BV, Boskalis BV)**
Project Manager: **Natuurmonumenten (society for the conservation of nature in the Netherlands)**
Area: **135 ha**
Other figures: **1000 ha of new nature**

The River Meuse comes from France and flows via Maastricht and Den Bosch through to the Dutch delta. The Meuse is a river with a pluvial regime, which means that there are large differences during the year in the supply of water, ranging from periods of drought to huge peaks. In Limburg in particular, the Meuse valley has really become too narrow to cope with the periods of high water. The 1990s saw major flooding in villages (Itteren and Borgharen) and town centres (Venlo and Roermond). The Meuse Works were developed to tackle the problem. This programme has three objectives: water safety, the development of new nature and the extraction of raw materials. Water safety is essentially being improved by widening and deepening the River Meuse. The revenue from the gravel, sand and clay that is extracted is used to partially fund the redevelopment of the areas (an approach known as self-funding). The Meuse Works have a budget of over half a billion euros.

Grensmaas and Zandmaas

The Meuse Works are subdivided into the Grensmaas and the Zandmaas, referring to two sections of the river. The Grensmaas (literally 'Border Meuse') forms the border with Belgium and is unnavigable north of Maastricht. Inland shipping can use the Juliana Canal that runs parallel to the Meuse. The Grensmaas river is being

Gravel processing sites in the construction harbour at Itteren

widened at twelve different locations, over a distance of 40 km. This is the largest river project in the Netherlands. Eleven subprojects are being executed by the Grensmaas consortium. This consortium was given a concession to extract 54 million tons of gravel and 10 million tons of sand. The Grensmaas project is the first in the Netherlands to combine protection from high water levels, nature development and the extraction of raw materials. A total of 100 million tons of soil is being moved, more than 12 km of dikes are being raised and the project will result in about 1000 ha of new nature. The Zandmaas stretch of river (literally 'Sand Meuse') starts at Maasbracht (near Roermond) and continues to Den Bosch. This is a distance of 150 km, along which a series of subprojects have been planned. The aim of the Zandmaas project is to provide protection against high water levels as soon as possible.

Grensmaas pilot project

A pilot for the Grensmaas project was started before the turn of the century in the vicinity of the village of Meers. 'Meers' literally means marshland or low-lying pastureland in Dutch. As the Romans used to say, *'Nomen est omen'* – the name is a sign. Three quarters of this little village of 1,300 people on the east bank of the Meuse was submerged during the floods of the 1990s. Trees were washed away and entire sections of the riverbank were broken off. In 1996, new dikes were built around the village, which is surrounded on three sides by the River Meuse. The pilot for the Grensmaas operation started two years later and continued until 2006. In the follow-up phase, the project area was increased from 46 to 135 ha.

Shallow gravel extraction

The pilot project was the first time that high water protection (along a 2 km section of river) had been combined with gravel extraction and nature development. The gravel was extracted in shallow waters wherever possible to create a much wider natural river. This in turn forged optimum conditions for new nature. Thus the area around Meers gave an indication on a small scale of what the entire Grensmaas river landscape would eventually look like. Over the past few years, many tons of gravel have been transported from Meers to the harbour in nearby Juliana Canal. Deeper gravel pits are being filled with clay from the Grensmaas sites of Urmond and Maasband. Later, soil from outside the construction area may be used. Meers will remain the setting for gravel extraction and nature development until 2024.

Meers is discovering what the Meuse has to offer

The new nature reserve near Meers is part of the RivierPark Maasvallei area. The idea is that an attractive, cross-border country park will be created, eventually covering 2500 ha. No hiking paths or routes have been laid out in the country-side around Meers; walkers are free to roam wherever they wish. Forest wardens employed by Natuurmonumenten and local residents from Meers worked together to decide on the best spots for access gates. The village has had a love-hate relationship with the Meuse for centuries. But now that there is so much to enjoy and villagers no

Increased rate of water flow in the Meuse at Meers; the lake on the right is a sand extraction site

longer need to fear flooding, the love side is gaining the upper hand. Or as the Meers Village Development Plan puts it: 'We want to rediscover the Meuse and make use of its potential by working to support the enjoyment of nature by local residents and visitors.'

Spectacular nature in and around the wild water river

The Meuse near Meers lives up to its reputation as the Netherlands' only wild water river. The sudden drop, caused by a geological barrier in the river bed, produces a spectacular increase in the speed at which the water flows. Gravel ridges that have grown spontaneously, gullies caused by erosion and a little peninsula covered with riparian woodland make for a popular destination for canoeists. The gravel banks purify the river water in a natural manner. This explains why the place is packed with waterfowl. Freshwater fish such as barbel and chub come here to mate. The higher-lying gravel banks are a fine sight when they are overgrown with wild marjoram and thyme. Plant seeds are carried from the south by the Meuse. The arrival of beavers from Germany after an absence of two centuries was a notable event. After another thirty beavers were released into the wild, the Meuse valley had a viable population. There are now an estimated 500 beavers. They play an important role in the nature management system as their gnawing and building activities introduce variety in the habitat, which in turn enables great diversity in the species present.

The rapid in the Meuse near Meers

The nature reserve near Meers is criss-crossed by attractive rambling paths

Woodhenge landscape art

Some fossilised trees were unearthed during gravel excavations near Meers, as happened in Millingerwaard (see p. 230). Fifteen trees were found in total. The oldest turned out to date from as long ago as 2425 BC. The other petrified giant trees are between 1400 and 1700 years old, making them early medieval in origin. In 2006, they were placed in a 'magic circle' in the Meerser Koegriend, a field next to the Meuse, as a monument commemorating the return to nature. Like Stonehenge in England, this Woodhenge functions as a calendar, as is evident from a spherical Meuse boulder placed in the middle of the circle. When the sun reaches its highest point on the summer solstice and winter solstice, the shadows of the tops of two trees fall precisely on this boulder. Two other trees mark the spring and autumn equinox. The direction of the wind and the movement of the sun across the sky are also registered. Six trees stand in a line with the villages in the surrounding area while all the trees in combination have been laid out to represent the path of the meandering River Meuse.

'Woodhenge' in the Meerser Koegriend along the Meuse

Sample Sheet of Intervention – overview on page 24-27

From village to island

Programme: **Grensmaas Meuse Works**
Province: **Limburg**
Coordinates **50°53'07.1"N 5°41'43.6"E**
GPS: **50.885292, 5.695433**
Assignment: **reduce the high water level**
Implementation: **2010-2014**
Water board: **Limburg Water Board**
Design: **Directorate-General for Public Works and Water Management in collaboration with the provincial authority of Limburg**
Contractor: **Grensmaas consortium: contractors united in Kombinatie Regenboog (Van den Biggelaar Aannemingsbedrijf BV, Van Oord Nederland BV, Boskalis BV), nine gravel companies united in Geo-Control BV**
Project Manager: **Staatsbosbeheer (nature management agency), ARK Natuurontwikkeling**
Fall in water level: **60-80 cm**
Other figures: **125 ha of new nature, 3.5 km of ring dike**

Borgharen, a village with a population of over 1800 just north of Maastricht, has been hit by flooding a number of times over the past few decades. Persistent heavy rain led to flooding in streets, cellars and access roads, and some of the local residents had to be evacuated. For example, just before Christmas 1993 the river Meuse reached a record level of 45.9 m above Amsterdam Ordnance Datum. Under normal conditions, the level in winter is almost 5 m less. And while the average river discharge at this point was usually 250 m^3 per second, discharge rates were more than ten times higher during periods of major flooding. Back in 1996, a 3 km ring dike was built around the village, providing protection against high water levels with an annual risk of 1 in 50. Then, as part of the Grensmaas projects, the channel was widened over a length of 3 km and the floodplains lowered. These measures improved safety levels to an annual risk of 1 in 250.

Borgharen was the first of the Grensmaas project locations to see the work completed, in 2014. Sand and gravel were transported to the gravel pit to the north of the village. The final layer of this topsoil depositary was filled using soil excavated during the construction of the A2 motorway tunnel under Maastricht, which opened at the end of 2016. The desired level of protection was achieved through the combination of the river widening and the dike.

Island

Due to the construction of the ring dike, the village now becomes an island when water levels rise. Under normal conditions, the greenery-covered elevation in the landscape is a popular destination for walkers and joggers. Once the Meuse's rate of discharge reaches 2280 m^3 per second, residents park their cars on the opposite side of the Juliana Canal to the east and a shuttle service starts. If water levels rise further still, this service stops

14

18

Foraging in wild nature

A Borgharen resident tries to pump water out of his house, 1993

The Grensmaas overflows its banks before the interventions, 2010

Borgharen, isolated by high water, 2003

Major news story: discovery of a centuries-old horse grave

A remarkable discovery was made at the Borgharen works site in 2010. In a ditch measuring 2 x 50 m, archaeologists found a grave with the skeletons of 65 horses. The horses are thought to have died during the French siege of Maastricht in 1794. Never before had so many horses been found in a single grave in Western Europe. An artwork made of gravel blocks and polished stones was placed on the site of the discovery.

The Brabant artist Adri Verhoeven has reproduced the life-size silhouette of the horse burial site in the landscape

The horse burial site at Borgharen reveals its secrets, 2010

and residents are evacuated. The dike would still provide sufficient protection even then, but the village would become completely cut off. From 2014 onwards, Borgharen's residents immediately started to notice that the water was not rising nearly as far as it used to whenever there were periods with heavy rain. A measurement of the interim result in 2016 confirmed their impression: the high water level had indeed fallen by 60 to 80 cm thanks to the river widening.

Unusual flora and fauna feel at home

Not only has the risk of flooding been reduced by a factor of five, Borgharen has also acquired 125 ha of new nature. The excavations turned the area upside down. Now a lovely mosaic landscape has been created with grasslands full of flowers, willow coppices and flowering bushes. Wild Polish primitive horses and Galloway cattle play a key role in the management of the nature. Together with beavers and deer, they prevent the area from becoming overgrown and create room for unusual flora and fauna. Ospreys, storks, Arctic terns and cranes pause here for a rest during migration. The river is home to seagulls, ducks, geese, swans and waders. Whitethroats and stonechats can be heard singing in the bushes. Sand martins dig out their nests in the precipitous sides of the riverbank.

Local residents design recreational routes

Borgharen's inhabitants are very enthusiastic about their new nature area. They came up with their own ideas for hiking trails and cycle paths. Dozens of locals have been trained as nature guides. When water levels in the Meuse are low enough, you can explore the river bed. Higher up are paths for roaming along that were created naturally by grazing animals and walkers. Stepping stones keep your feet dry in marshy spots. Better connections have been created between the cycle paths crisscrossing the Rivier-Park Maasvallei area.

Wild Konik horses and Galloway cattle prevent the vegetation from growing too thickly. Belgium is across the water

Sample Sheet of Intervention – overview on page 24-27

Gravel depositary, nature area and water storage

Programme: **Grensmaas Meuse Works**
Province: **Limburg**
Coordinates **50°57'43.9"N 5°43'53.9"E**
GPS: **50.962190, 5.731647**
Assignment: **reduce the high water level**
Implementation: **2015-2024**
Water board: **Limburg Water Board**
Design: **Directorate-General for Public Works and Water Management in collaboration with the provincial authority of Limburg**
Contractor: **Grensmaas Consortium: contractors united in Kombinatie Regenboog (Van den Biggelaar Aannemingsbedrijf BV, Van Oord Nederland BV, Boskalis BV), eleven gravel producers united in Geo-Control BV, Natuurmonumenten (society for the preservation of nature in the Netherlands)**
Project Manager: **Natuurmonumenten**
Drop in water level: **up to 110 cm**
Area: **1075 ha**
Earthworks:
Other figures: **54 million m³ of gravel extracted**

Rivers started transporting large quantities of gravel to what is now the Netherlands in the early Quaternary Period, about 2.5 million years ago. The ancient pebbles – remnants from almost all periods in the Earth's history – were deposited in the lower lying parts of the river valleys. The thick layers of gravel on the river bed and under the floodplains of the Meuse in Limburg originally came from the Vosges in northern France and the Ardennes in Belgium. After the Second World War, Limburg became the scene of large-scale gravel extraction. The province became the country's leading supplier of gravel.

10

14

Ruining the landscape or paving the way for nature development?

All that digging resulted in large, deep pools dotting the Limburg countryside with very little biodiversity. By the end of the twentieth century, the local population and nature organisations had had enough of this. Farmers also started to protest about the gravel extraction as farmland was becoming increasingly scarce. Back in 1989, the provincial authority of Limburg decided to phase out large-scale gravel extraction over the period to 2025, using a sustainable approach in which the large, deep pools would make way for 1100 ha of nature development and the restoration of the River Meuse's ecology. From 1989,

The village of Visserweert remains accessible during high water thanks to a new bridge

economic and ecological concerns went hand in hand; flood prevention was added as an objective after 1995. The Meuse Works were able to benefit from a development that had already been set in motion.

Koeweide

The water storage facility with the largest capacity in the Grensmaas project is being built near Koeweide in the municipality of Sittard-Geleen. The channel is being widened by several hundred metres. A second safety dike is being built further inland on the southern bank where the Meuse makes a sweeping bend. That will create more room in this bottleneck too. The rest of the dike will remain in place because of its great cultural and historical significance.

A good 35,000 shiploads

Even more is going on now at the Koeweide location. In autumn 2015, a works harbour was built at the site of the former municipal rubbish tip Trierveld/Koeweide and temporary storage space was created for 200,000 tons of dredged gravel from the works sites of Nattenhoven, Grevenbicht, Visserweert and Trierveld/Koeweide itself. After the 'Rotterdam' processing plant arrived at the works harbour in March 2017, gravel processing could start. A second processing plant, 'Merwede', was added later. In the period to 2025, a good 35,000 ships will be setting off from here, carrying a total of 54 million m^3 of gravel and sand to buyers throughout the country. To facilitate this, an open passage was dug between the harbour and the nearby Juliana Canal.

Topsoil depositaries will have function as nature and farmland

The river widening activities at the different work locations produce not just gravel but also unsellable topsoil (dredge spoils). This is being deposited in the deep gravel pits that are left behind in the Trierveld/Koeweide and Koeweide-West sites. Filling in these pits creates the basis for their future functions. This is already being taken into account when deciding what height the future ground level should be. The topsoil depositary covering 40 ha in Trierveld/Koeweide will eventually become farmland. Because of the relatively high water levels, the soil must not be too wet and that means the area must not be too low lying when it is delivered at the end of the project. The 36 ha topsoil depositary in Koeweide-West will be turned into a nature area. It will take the form of a shallow gully. Once the gravel extraction and river widening activities have finished, the Grensmaas consortium will hand the sites over to Natuurmonumenten.

The infrastructure has been adapted to ensure that sand transportation causes a minimum of inconvenience

Impressive river widening of the Meuse. Above right is the bridge across the future channel, near the village of Visserweert

Sample Sheet of Intervention – overview on page 24-27

10
13
14
16

New insights give big boost to flood protection and the quality of the area

Programme: **Zandmaas Meuse Works**
Province: **Limburg**
Coordinates: **51°26'10.6"N 6°09'52.0"E**
GPS: **51.436283, 6.164435**
Assignment: **reduce the high water level**
Implementation: **2006-2015**
Water board: **Limburg Water Board**
Design: **Directorate-General for Public Works and Water Management in collaboration with the provincial authority of Limburg**
Contractor: **Delfstoffen Combinatie Maasdal Exploitatie Lomm BV**
Project Management: **The Limburg Landscape Foundation**
Drop in water level: **12 cm**
Area: **99 ha (39 ha for the channel itself, 60 ha of nature)**
Earthworks: **8.2 million m³**
Other figures: **length of high water channel: 2 km**

Lomm is a village on the Zandmaas section of the River Meuse, located about 10 km north of Venlo. On the east side of the Meuse, Delfstoffen Combinatie Maasdal Exploitatie Lomm BV (DCM) is building a high water channel with a length of 2 km. When water levels rise, the channel flows with the Meuse, thereby helping the discharge of water. This has led to a noticeable drop in water levels in Venlo. While carrying out the work, DCM saw an opportunity to reduce water levels further than was envisaged in the design contained in the Tracébesluit Zanmaas/Maasroute, theZandmaas/Meuse Route Transport Infrastructure Decree (2002). What is more, making the river even wider created benefits for nature, the landscape and recreation. In 2013, the go-ahead was given for the proposed changes.

Reduction in floodplain land levels

In the original design, the dimensions of the channel were based on a high water level with an annual risk of 1 in 250 and with water discharge of 3275 m³ per second. This enabled a reduction in water levels of 7 cm to be achieved. By lowering even more floodplains to the east of the channel, it turned out to be possible to reduce the Meuse water levels by a further 5 cm, giving a total reduction of 12 cm. The high water channel is now designed for an even greater discharge rate of 3830 m³ per second, with an annual risk of 1 in 1250. This adds 11.4 cm to the reduction in

A suction dredger and a floating sand classification installation have started extracting and processing industrial sand from the initial extraction site, 2010

Large machines were deployed to build the high water channel still under construction, 2006

water levels. As a result, the contribution to the high water target for 2015 at Lomm has been achieved.

Bypass relieves pressure

Permission to widen the river even further was, however, subject to one condition. While this alteration would reduce water levels upstream, it would create an additional peak in Meuse river levels downstream. This undesirable effect, which of course could not be allowed, has been offset by adding a bypass to the north of the channel along which any additional water is diverted. A nice side effect is that the river has been given even more room thanks to the bypass.

More excavation and sales of sand and gravel

An important consideration for DCM was that the expansion of the area from 80 to 90 ha allowed it to extract more raw materials. This is because the construction of the high water channel is a so-called self-funding project: the idea is that the costs should largely be covered by the extraction and sale of sand (80%) and gravel (20%) for industrial purposes. The excavated materials are used in cement factories, asphalt plants and housebuilding. However, DCM has to stick to a 'balanced cut and fill' arrangement, keeping to proportions of over 60% land and just under 40% water in the high water channel area. This means that not all the raw material present in the area can actually be excavated. The original design allowed for the extraction of a maximum of 5.4 million tons of sand and gravel. An additional 3.6 million tons could be added to this thanks to the further widening of the river. Of course, this too has to comply with the balanced cut and fill arrangement.

Greenhouses make way for nature

Another advantage of the amended design is that the area with market-garden greenhouses in the south east can get a quality boost. The enlargement of the channel necessitated moving two greenhouses and rerouting a 1 km long embankment. A more attractive transition in the landscape has been created by setting up the 7 ha area for nature development and extensive recreation after excavations have finished. A second argument for the removal of the greenhouses was that they were located in the flow area of the River Meuse. The greenhouse owners hardly had any room to expand because they are subject to the strict rules of the Major Rivers Policy. Moreover, whenever water levels rose, the greenhouses exerted considerable upward pressure on the water in the Meuse.

Villagers want to unwind along and on the Meuse

The bypass to the north of the high water channel connects up the village of Lomm, which fits in nicely with the wish of local residents to have more links between the village and the Meuse. DCM invited the Lomm village council to help come up with ideas for an integrated area plan with room for hiking trails, cycle paths, gentle water sports and sunbathing lawns. Other parts of the area will have a dedicated nature function with limited recreational opportunities, and will be managed by the Limburgs Landschap Foundation.

A barge makes its way over the Meuse to the left of the high water channel

Rituals at a place of worship

ADC ArcheoProjecten was commissioned by DCM Exploitatie Lomm to conduct a large-scale archaeological survey in the vicinity of the high water channel. When digging in a field right next to the Meuse, the archaeologists found a grave field and a place of worship dating from the Iron Age (c. 800 BC to 12 BC). The grave field contains 60 graves, making it one of the largest in the Netherlands from the Late Iron Age. The place of worship is older. It consists of four trenches that form a rectangle with sides of 33 m and 37 m. Within it is a second, smaller rectangle measuring 7 x 8 m and also enclosed by trenches. There may have been a small temple on the site too. Archaeologists suspect that rituals were performed here in the Middle Iron Age. These are just some of the discoveries that were made here. The Rijksdienst voor het Culturele Erfgoed (Cultural Heritage Agency) of the Netherlands has designated the site an area of very high archaeological value.

The Iron Age cult site, 2008

Spatial Quality

The water protection programmes owe their success to the fact that safety is linked to quality. That was literally the case in the 'Room for the River' programme, which had the dual objectives of water safety and spatial quality. The latter is interpreted as the value of an area in terms of its utilisation, people's experience of it and what it offers for the future. The underlying idea is that the area traversed by the main rivers is of international significance for the economy, the ecology and the landscape. This significance demands and justifies the effort expended on spatial quality when making interventions. In particular, this means that the economy, nature, recreation and water safety are seen in conjunction with one another. The dual objectives were evident in the quality of the design and the combination with other spatial programmes. But the positive side effects were possibly even more interesting: broad support among local residents and businesses, the involvement of public administrators at all levels and an international reputation. Spatial quality boosts the appeal of a place for potential new residents and businesses.

On the Agenda

'Room for the River' was followed by the Flood Protection Programme, with its slogan of 'sober and effective'. According to Dutch Special Envoy for International Water Affairs Henk Ovink, this does not have to stand in the way of good solutions. In the Netherlands, tackling water safety is unthinkable without public participation and an integrated approach in which different interests are taken into account. Spatial quality is the trump card used to get affected residents and businesses on board. The problems only arise when spatial quality is seen as an optional luxury item. Ovink: 'If we focus exclusively on safety, then we will lose the battle. It will only work if we have safety *and* quality, with an inclusive, design-driven process in which an integrated long-term approach leads directly to integrated projects in the here and now, and in which that mix is continuously being made. Inclusive, integrated, long-term, short-term, innovative and transparent so that we can learn and be held accountable. That seems complex but it is simply necessary and a huge opportunity! This is our great strength in the Netherlands.'

Guide, Knowledge Team and Quality Team

Spatial quality as an objective is a good starting point. But who decides whether the quality of a development plan or design is satisfactory? What criteria should be used for this? The publication of the Spatial Quality Guides for the Waal, Rhine and IJssel rivers meant there were now reference documents describing the underlying logic and distinctive features of each branch of the rivers. Designers can use this information as input. It helps them preserve and strengthen the characteristics of the river landscape specific to that area. A second way to make work of spatial quality was the establishment of a Spatial Quality expert team by the Directorate-General for Public Works and Water Management. Spatial quality is something that is difficult to express in 'hard' quantitative data. It is a question of tailored solutions, dialogue with the affected parties and ultimately an independent assessment (with recommendations). To achieve this, a quality team was set up, consisting of a specialist in physical geography, an urban planner, an expert on rivers, an ecologist and a landscape architect.

The Role of the Quality Team

The quality team's contribution to the 34 projects in the 'Room for the River' programme was tightly organised. In each project phase, the quality team held discussions with the project team to agree on their goals for spatial quality and how these should be achieved. Location visits were organised in the feasibility study phase and three formal advisory reports were produced for each project. That led to both minor and radical changes to the design. For example, in Westenholte in Zwolle (see p. 248) the dike was set back further than had been envisaged in the original plan. Other adjustments were made in the Avelingen business park, Natuurderij country estate in Deventer, clay and sand excavation in Millingerwaard and Huissensche Waarden, the use of the Overdiepse Polder for agriculture and in Kampen town centre (see p. 242, 230, 200, 132).

Positive Feedback

The establishment of a quality team was one of the innovations of the 'Room for the River' programme. The team could reckon on considerable support for its input, both from local residents and from the key figures working on the projects. The independent nature and substantive content of the team's advice meant that its recommendations were respected by all parties. The attention paid to quality throughout the process made it possible to achieve quality improvements in a relatively straightforward manner in the projects. Because this was done so promptly, delays later on could be avoided. Not only did the quality team safeguard the objective of spatial quality, they also helped make sure the programme was implemented smoothly.

Systematic Approach

The positive experience with the 'Room for the River' quality team led to a follow-up for the Afsluitdijk causeway. The Flood Protection Programmes are having to make do without a dedicated quality team. Even so, spatial quality remains an important theme because of the nature of the projects: the involvement of local residents and businesses is essential and cultural heritage objects are often affected. It will be interesting to see where this approach leads. PhD candidate Ties Rijcken in 2017: 'The water sector has built up statutory systems for safety, water quality and shipping that are based on cycles of studies, assessments, measures and more assessments to check whether the intended results have been achieved. Such an approach is more difficult and less developed for nature and landscape quality, but it is certainly possible. You could have a national quality team give an appraisal of the spatial quality of the landscape adjoining areas of water, in addition to the national dike and water quality assessments. Just don't call it an assessment; call it a tool for a systematic approach to spatial quality that can help raise the standard of the debate on this subject.'

Relaxing on Scheveningen's new boulevard

The Lentse Warande

Design Quality

The demand for a national barometer is legitimate. The results will only become evident in the actual projects. All the projects that have been carried out over the past few years have enriched the Netherlands by adding a variety of clearly articulated, unusual landscapes. The design quality is reflected in the architecture, the landscape, the utilisation and people's experience of the space. A wide range of civil engineering structures were created: bridges, culverts, locks and sluices, pumping stations and dikes. Outmoded, run-down areas were modernised, sometimes with spectacular results. Scheveningen needed rather more allure, and got it when the boulevard was revamped. The beach at Katwijk has become more appealing thanks to the easy access offered by the drive-in dunes. Extensive parks and nature reserves have been created next to the rivers. They have not only improved nature values and water safety but, most importantly, they have boosted the quality of life in the Netherlands as a whole. In towns such as Nijmegen, Venlo and Deventer, the water safety assignments gave them a dose of urban style, plus areas that have enhanced the positive image of these towns. The best proof of the successful integration of the water safety projects and spatial quality is the completely natural manner in which the projects have been absorbed into the Dutch landscape.

An elegant bridge links the shores at the new marina in Cadzand-Bad

Enjoying the magnificent Noordwaard landscape

Urban Development and Protected Views

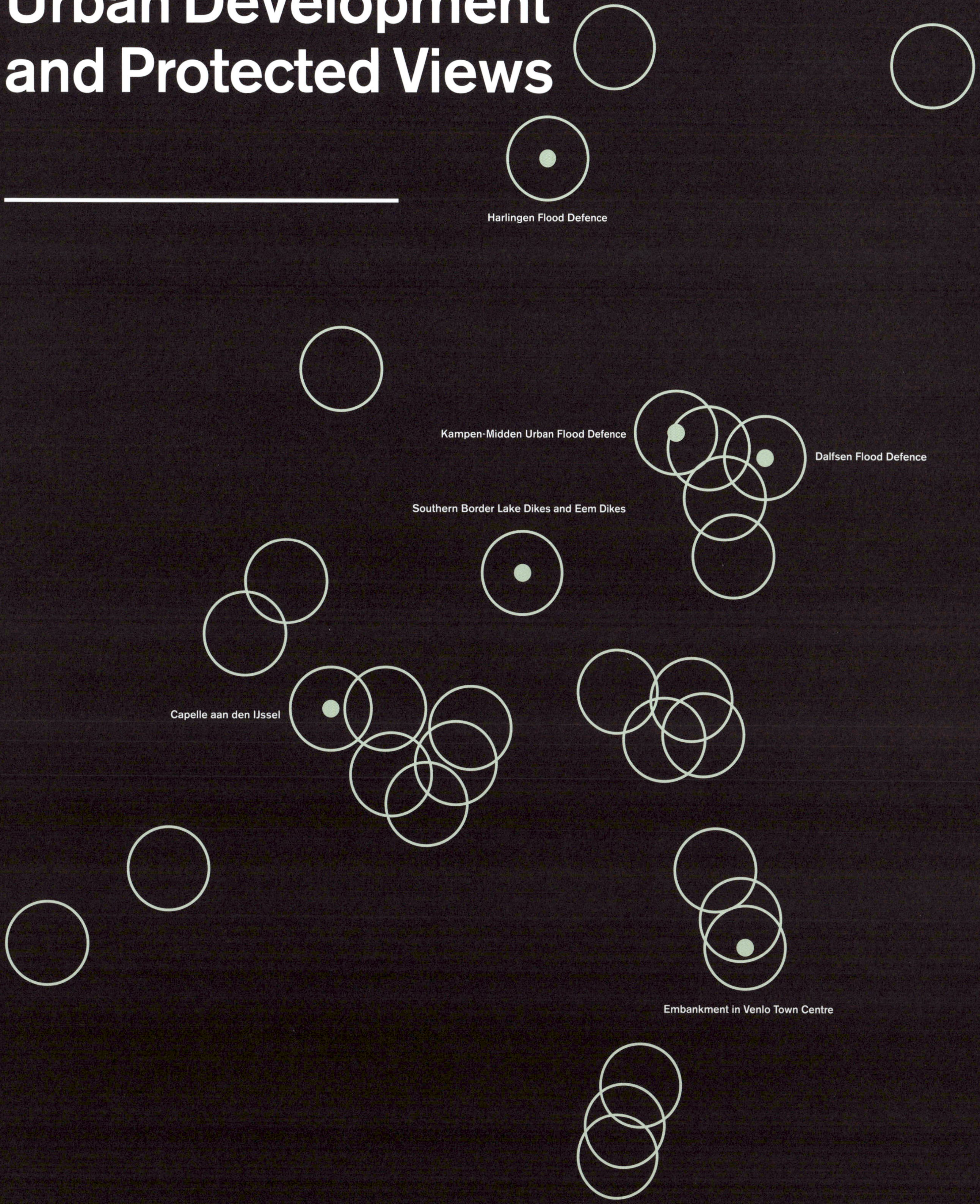

Flood Defence Harlingen – Looking out over the sea from the robust harbour wall, 2017

Embankment in Venlo Town Centre – Restructuring and new construction have transformed the once run-down Meuse river front in Venlo into a vibrant part of the city, 2015

Kampen-Midden Urban Flood Defence – Resting after hoisting a lifting gate during the annual emergency drill, 2010

Voorstraat
HOOG WATER BRIGADE
HOOG WATER BRIGADE

Sample Sheet of Intervention – overview on page 24-27
03
05
08

Tailored solution on the Meuse

Programme: **Zandmaas Meuse Works**
Province: **Limburg**
Coordinates: **51°22'10.6"N 6°09'53.0"E**
GPS: **51.369616, 6.164708**
Assignment: **protection against flooding behind the quayside with 1/250 annual risk**
Implementation: **2002-2006**
Water board: **Limburg Water Board**
Design: **Directorate-General for Public Works and Water Management, Meuse Works**
Contractor: **GMB/Oldenkamp**
Project Manager: **Limburg Water Board**
Length: **2 km**

Towns and villages on rivers are tricky problems in flood protection programmes. The embankment and waterfronts often go back centuries and do not meet modern-day requirements. If anything should go wrong, large numbers of people are affected and the economic damage can be substantial. But the town or village often literally stands in the way of a solution: there is no room to widen the dikes, little public support for raising the dikes and no space to widen the path of the river. How can all those bottlenecks be dealt with? A wide range of tailored solutions have emerged, often working with customised fixes for the old village and town centres with raised embankment and demountable barriers. Just outside those centres, the aim is to create urban parks with abundant water, which enrich the towns and offer a safe place for storing water and secondary water-courses.

Raising the embankment in an old town by 1.5 metres

Venlo town centre is a notorious bottleneck for the River Meuse. When water levels rose in 1993 and 1995, the embankment flooded and many streets became impass-able in no time. Water levels remained high for several days on both occasions. Sandbags stacked in front of the doors had little effect. After the second disaster in 1995, it was clear as daylight to every-one that urgent measures would have to be taken in both Venlo and Roermond,

The streets of Venlo were flooded due to the high water level of the Meuse in 1995

Cars are parked on the quays along the Meuse when the water is low, 2011

Installing stop logs keeps the water out when the water is high

a town 25 km upstream with similar problems. It was possible to relieve Venlo town centre to a large extent by widening the river and creating secondary channels in the surrounding area. Ways of raising the embankment in the Meuse town itself were considered. Calculations showed that the embankment would have to be raised by 1 to 1.5 metres in order to protect Venlo against extremely high water levels. But how could such a project be carried out?

Implementation in consultation with the people immediately affected

Two issues were determining factors in the detailed plan for the embankment: residents living next to the Meuse understandably did not want their view of the river obstructed by a high wall, and the town administration wanted to strengthen the town's relationship with 'Mother Meuse'. For decades, the town centre had faced away from the water. Now the intention was to take the river as the starting point for urban development in that area. A customised solution was needed. The precise details of parts of the quay plan were decided in close consultation with local residents, town administrators and other stakeholders.

Demountable embankment for good views

In 1996, the water boards had built emergency dikes wherever there was room to do so. They provided protection against high water levels with an annual risk of 1 in 50. As part of the Meuse Works, the decision was taken to improve safety further still and reduce the annual risk of flooding to 1 in 250. That meant the embankment would have to be raised in Venlo. The first raised embankment in Venlo were delivered at the end of 2005. They largely consisted of dikes constructed from clay. As similar projects had been completed in the same period in Roermond, Gennep, Mook and Middelaar, the risk of flooding had now been reduced to the new norm for 70 to 80 per cent of the population living along the River Meuse. The remaining dikes and quay walls were reinforced from 2006 onwards. Demountable embankment were used in places where good views of the Meuse were important. As soon as high water levels seem imminent, the water board can raise low brick walls with partitions to the appropriate height.

Facing the River Meuse

In the vicinity of Venlo's historic town centre, embankment measures have been integrated into the urban residential area that has been created here since 2006. Redevelopment and new buildings have transformed what was once a run-down waterfront area on the Meuse into a lively district. The quay wall made from *Petit Granit* (a Belgian limestone) and stretching for 1 km marks the old town. Shops and apartments have been built around an inviting short-stay marina. People stroll along the boulevard and stop at the pavement cafes. Broad steps lead down to the lowest level of the quay wall, which functions rather like stands overlooking the water. The natural floodplains between the town and the Meuse have been laid out as a park (Maaspark). The grass slopes demarcated by basalt boulders make a pleasant spot to relax in the summer months. As the season changes and water levels rise, the river can take over again. Venlo is once again facing the River Meuse.

During high water in 2011 the collapsible flood defence was used at Blerick

The steps descending to the water in Venlo's marina

Venlo's marina

New bridge over the
marina to the bank of
the Meuse near Venlo

Romeinenweerd: from clay pit to nature area

The material required for the construction of the emergency dikes in 1996 did not have to come far. Clay could be excavated in the pastureland of Romeinenweerd, just south of Venlo, on the west bank of the Meuse. Crane operators took clay from various spots. The lack of a planned approach resulted in a disfigured landscape with three large pools and a few smaller ones. But this was soon recognised as an opportunity. If the area was left to run wild, the original river flora and fauna would be able to return. After only a couple of years, surveys were already finding 300 plant species and numerous rare birds. A unique feature of the pools is that the surrounding groundwater seeps to the surface here. As a result, the area is now home to some unusual marsh species, often rare plants that thrive in a habitat with seepage water rich in iron, such as the marsh marigold, the water violet, the water horsetail and the bog beacon. When water levels are high, the Meuse floods the Romeinenweerd area.

The remarkable development in the nature in Romeinenweerd led to it being designated a model project for the Meuse Corridor project. This project has been running since 1999 and is an initiative by four municipalities, the provincial authority of Limburg, the World Wide Fund for Nature, the Limburgs Landschap Foundation, Staatsbosbeheer, the Peel and Maasvallei water board (now the Limburg water board) and the Directorate-General for Public Works and Water Management. The idea is that lasting protection against high water levels should go hand in hand with attractive areas for tourism and recreation and with closer links between the Meuse and the adjoining towns and villages. Following the example of Romeinenweerd, other former agricultural areas along the Meuse in northern Limburg have been returned to nature. A 25 km long ribbon of marshes, gullies, grasslands filled with flowers, brushwood and riparian woodlands has now been created. By only building dikes where absolutely necessary and giving the river free rein outside those areas, it was possible to leave river beds and channels, dating back thousands of years in some cases, almost completely intact in the landscape.

North of Venlo these sheep take charge of nature conservation along the Maaskade

The Romeinerweerd was created by works carried out after the extremely high water level in the Meuse in the mid-1990s

Sample Sheet of Intervention – overview on page 24-27

Final piece in Delta Works puzzle

Programme: **Delta Plan**
Province: **Friesland**
Coordinates: **50°58'00.2"N 5°44'12.1"E**
GPS: **50.966712, 5.736703**
Assignment: **dike reinforcement**
Implementation: **study and planning phase 2002-2007, construction 2008-2010**
Water board: **Wetterskip Fryslân**
Design: **Royal Haskoning, Wetterskip Fryslân, (structural)**
Contractor: **Consortium of Van den Herik Sliedrecht and Jansma Drachten**
Project Manager: **Wetterskip Fryslân**
Length: **500 m new water defence structure, 6.75+ Amsterdam Ordnance Datum**
Other figures: **4 m wide pedestrian promenade surmounting the defence**

Harlingen is a port town on the Wadden Sea in the northern Netherlands. It is one of the eleven old towns of the province of Friesland. Harlingen's new flood defence officially became operational on 24 August 2010. The Delta Plan that had started in 1957 was now finally complete. It was almost 60 years since the great North Sea Flood, which had prompted the Netherlands to build this massive series of coastal defences . But why had it taken so long after the completion of the penultimate Delta structure – the Maeslant barrier near Rotterdam in 1997 – to improve flood protection in the Frisian seaport? As long as there were still plans to reclaim the Wadden Sea (one day), there was no urgent need to raise the Wadden dikes. Only when this idea was dropped for good were the Wadden Sea dikes raised to the height required by the Delta Plan. The water defences in Harlingen were the final piece in the puzzle.

A complex hub

Wetterskip Fryslân faced a testing challenge as everything converges on this small section of the Wadden coast at Willemshaven harbour: trains, buses and ferries to the islands of Vlieland and Terschelling, plus thousands of cars, cyclists and pedestrians in the holidays and on sunny weekends. How can you take account of the wish for more traffic safety, greater access to this dynamic area and closer links between the mudflats and the town in the design for the water defence system? The solution was surprisingly simple. The key was to move the retaining wall so that the railway line and platform were inside the dikes rather than outside as before.

This keyhole provides a glimpse of the city. During high water the opening in this retaining wall is closed off

Priority given to coordination with local stakeholders

That key was found after various possible routes had been subjected to an environmental impact assessment. All stakeholders were kept informed and consulted on a regular basis, in particular local residents, shipping companies and the railway company NS. This coordination guided the design process for the flood defence and the facelift for the surrounding area.

Sleek and unifying

The chosen route leaves plenty of room for the development of residential and other accommodation functions in the area around the Willemshaven harbour. Local residents are not affected by the shadows cast by the higher wall as it has been placed further away from them. The sleek structure means that there is no longer a bulky partition separating the town from the mudflats. On the contrary, the wall has brought the two closer together, for example through the 4 m wide Wadden Promenade that was built on top of the barrier and which gives views out over the Wadden Sea, the harbour and the historic town centre. Cyclists, pedestrians and emergency services can enter the harbour area through an opening in the barrier that looks rather like a keyhole. As soon as there is a threat of high water levels, the hydraulic sluice gate in the passageway is closed. Walkers are steered along access ramps that connect the railway platforms to the ferry terminal and the pedestrian promenade on the barrier. Halfway along the barrier, two footbridges cross the railway line towards the town centre. A third footbridge takes pedestrians to the embanknment. The fluid shapes of these bridges are reminiscent of a huge wave crashing over the sea barrier.

Ready for a second Delta Plan

With its new crest level of 6.75 m above Amsterdam Ordnance Datum, the 500 m long flood defence will guarantee protection for the next 100 years. The barrier is designed to cope with a rise in sea levels of 65 cm. That is 35 cm more than the norm that was used in the first Delta Plan. This was in anticipation of the new standards incorporated in the Water Act (2016), which contains amended forecasts for the rise in sea levels together with the associated norms.

Flexible flood defence gives slightly

The crest height was determined on the basis of both water levels and wave overtopping during a storm surge. The sea piers on the north and south sides of the harbour, which were also reinforced, have helped substantially reduce the force with which the waves crash against the flood defence. The overhang on the mudflat side of the pedestrian promenade also has the effect of weakening the power of the waves. Yet despite these two measures reducing the force of the waves, tests showed that the concrete flood defence would still be hit hard. It would need sound foundations – that is to say, sturdy but in particular not too rigid. On the land side, the flood defence is supported by dam walls that extend 12 m underground. On the sea side, the foundations consist of a row of tie bar poles extending 35 m under the ground. This makes the structure somewhat flexible. The horizontal reactive force from the ground is used to absorb some of the force of the waves. Thanks to the tie bar poles, the flood defence is able to move land-inwards by a few centimetres, as tall buildings do during a heavy storm.

The promenade provides a splendid view over the port and the Wadden Sea

Harlingen is protected against high water on the Wadden Sea

The tie bar poles enable the retaining wall to band land inwards by a couple of centimetres when waves crash against it during a storm surge

Reclaiming the Wadden Sea

Harlingen first acquired a water flood defence only in the 1960s. Until then, a storm surge could cause the seawater in the harbour to wash up over the embankment and railway line and flow into the Bolswardervaart canal. There were only two sluices to check the water. During the severe storm in 1956, a sea-faring vessel was laid on its side on the dock to absorb the shock of the waves. It was a close call. The water flood defence that was built in 1965 was a secondary structure. That was because a plan was being considered at the time to reclaim the land under the Wadden Sea between the island of Ameland and Frisian mainland. Harlingen would then become an inland harbour, and the water flood defence would therefore not need to be high. But in the 1970s, the land reclamation plan was shelved under pressure from environmental organisations. As a result, the defences along the Wadden Sea had to be raised to the height stipulated by the Delta Plan. That was finally achieved in 2010.

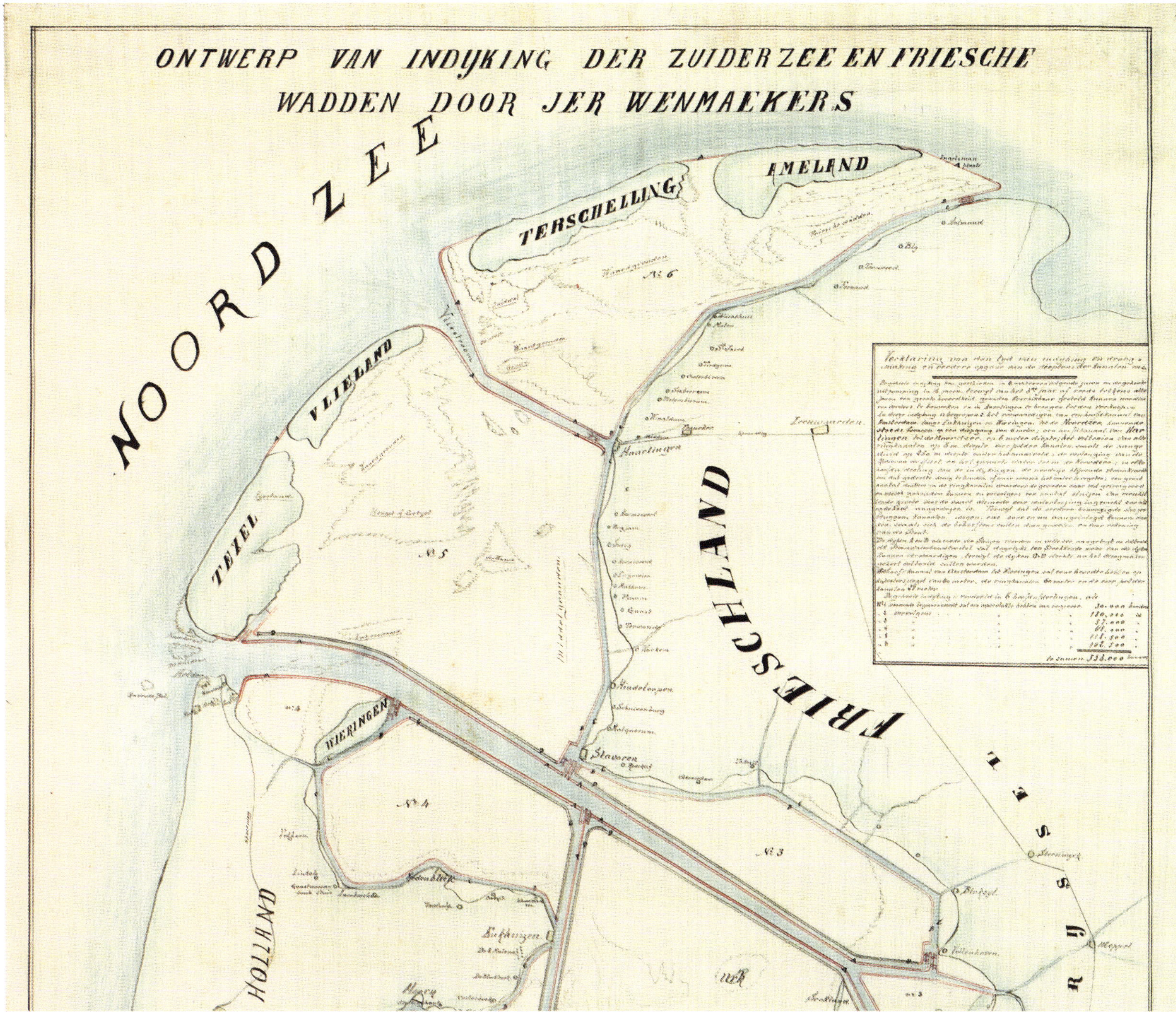

The plan of reclaiming large parts of the Wadden Sea was met by various protests and the plan was shelved in the 1970s

03
05
Sample Sheet of Intervention – overview on page 24-27

Room for a view of the village

With a length of 167 km, the River Vecht that runs through Overijssel is both the largest of the small Dutch rivers and the smallest of the major Dutch rivers. Climate experts predict that the river will be discharging 25 per cent more water by 2050 because of climate change. One of the locations that was not prepared for this was the village of Dalfsen on the north bank of the Vecht. When assessments were carried out in 2010, a 520 m stretch of the existing flood defence was found to be not up to standard. The village embankment turned out to be too low and the secondary channel opposite the village could not reduce water levels to a sufficient degree. Meanwhile, Dalfsen was already working on a major waterfront development project on this site. By combining dike reinforcement with urban development, it was possible to ensure that the new development complies with the new safety standards.

Development plans

The village of Dalfsen has always had a distinctive village skyline, with the bridge over the Vecht, the Westermolen windmill, some factories outside the dike and a small-scale waterfront behind the dike. The church towers above the rest of the village. The plans for the waterfront involved demolishing the old town hall and the Eshuis factory and constructing blocks of flats that contain residential accommodation, shops, restaurants and cafes, with a square and a pedestrian promenade along the River Vecht.
The people living in the Vecht valley had become estranged from the river in the course of the twentieth century as it had lost its function as a transport route – and consequently its economic importance. Dalfsen too had turned its back to the river. The municipal administration saw the riverfront development as the perfect opportunity to improve the quality of life and the local economy.

Programme: **High water protection programme / 'Room for the River Vecht'**
Province: **Overijssel**
Coordinates: **52°30'09.5"N 6°15'32.7"E**
GPS: **52.502649, 6.259073**
Assignment: **raise the dike and the quay**
Implementation: **2014-2015**
Water board: **Drents Overijsselse Delta water board (Groot Salland water board before 2016)**
Design: **RHDHV**
Contractor: **Dalfsen waterfront development consortium (OWD), ELJA Beton (embankment)**
Project Manager: **Drents Overijsselse Delta water board**
Length: **200 m long embankment (320 m long embankment on the east of the bridge will be tackled later)**
Earthworks: **550 m³**
Other figures: **barrier limit increased from 4.35 above Amsterdam Ordnance Datum to 4.70, to be increased in future to 5.40.**

In Dalfsen the local waterfront development was combined with the required dike reinforcement

Multi-level terraces

The water board had actually scheduled the work in Dalfsen for 2021. But it decided to bring the work on the 200 m section of embankment west of the bridge forward by seven years as the construction of a new flood defence could be combined perfectly with the waterfront development. The water barrier was raised by 70 to 80 cm. The difference in height was barely noticeable as it was concealed under the new housing blocks. Protected trees and two distinctive buildings – the chicory factory (built in 1858) and the white villa that was once the factory owner's home – were left standing between the new buildings and the bridge over the Vecht. Of course the embankment cannot be raised at the point where these buildings stand. The flood defence therefore takes the form of a wall along the middle of the embankment, with multi-level terraces down by the water and large steps over the wall. This produced an interesting public space that being inside or outside the dike something you can experience directly. Both buildings have been assigned a new designated use as hospitality facilities and the factory is also home to a theatre. The new town hall forms the western border of the riverfront. The building is at a safe height and faces the river. Dalfsen's new skyline is dominated by the higher lying volumes of the town hall and block of flats, the base with terracing and, situated behind this, the lower lying old buildings on the river.

Pre-financing essential

One consequence of bringing the construction of the water defence structure forward was that the water board had to pre-finance the project. The water board was, however, prepared to take this risk in the light of the expected gains in efficiency and quality. In September 2014, Dalfsen became the first project in the national Flood Protection Programme to get off to a start. What is more, Dalfsen has served as an example for similar projects elsewhere in the Netherlands thanks to its smart exploitation of combination opportunities.

Possible to raise levels later

The old flood defence in Dalfsen had a height of 4.35 above Amsterdam Ordnance Datum. That height has to be raised to 5.40 above Amsterdam Ordnance Datum to be sure of protecting Dalfsen from flooding for the next 100 years. But that is too much from the perspective of allowing people to experience the Vecht, which is an important guiding principle in the waterfront development. The wall has therefore been raised to 4.70 above Amsterdam Ordnance Datum. According to the water board, this height will be sufficient to keep Dalfsen dry for the next 50 years. The flood defence has been constructed in such a way that a mobile barrier in the form of a 70 cm high stop log can be placed on top later, if necessary.

Vibrating and pushing

It was feared that the process of inserting the dam wallboards that make up the flood defence into the ground would cause damage to trees and buildings. The contractor used the high-frequency vibration method for about half the dam walls, an approach that minimises noise nuisance. A pressure rig was used for the remaining walls, located near trees and buildings. The wallboards were pressed into the ground in 4 m sections and welded on to avoid vibration damage and damage to trees and branches. The contractor was able to give value for money by making only limited use of this expensive method. That was precisely the intention of the MEAT tendering procedure (in which the contract was granted to the Most Economically Advantageous Tender).

Dalfsen once again faces the river

A bird's eye view of the new waterfront in Dalfsen

Local residents make the most of an urgent dike reinforcement operation

Programme: **Flood protection programme**
Province: **South Holland**
Coordinates: **51°55'33.9"N 4°35'44.6"E**
GPS: **51.926077, 4.595732**
Assignment: **dike reinforcement**
Implementation: **2011-2016 (dike reinforcement completed seven months ahead of schedule)**
Water board: **Regional water authority of Schieland en de Krimpenerwaard**
Design: **Combinatie Mourik-Liebregts, Tuin- en landschapsarchitect Jos van de Lindeloof**
Contractor: **Mourik-Liebregts consortium**
Project Manager: **Regional water authority of Schieland en de Krimpenerwaard**
Length: **400 m inside the dike and 100 m outside**

The inhabitants of Capelle aan den IJssel in the province of South Holland got quite a shock when they learned that their dike was one of the weakest along the Hollandsche IJssel river. Not that they were afraid of getting wet, but how much of a subsidence risk would there be from the dike reinforcement work for properties located on or next to the primary flood defence? Would it have a negative impact on the appearance of the town, which has protected status, the two listed buildings or Slotpark, and the Slotpark, encircled by a moat and still showing the outlines of a thirteenth-century castle? And would the right measures be taken to safeguard access during execution and keep noise nuisance to a minimum? In 2011, the dike reinforcement work along the street of Dorpsstraat in Capelle aan den IJssel was designated a 'frontrunner project' in the extensive programme of dike reinforcements that the regional water authority of Schieland en de Krimpenerwaard is carrying out along the Hollandsche IJssel. The densely populated polders on either side of the river are liable to flooding and are also suffering from subsidence.

An operation with a major impact

The 400 m long stretch of dike along Dorpsstraat in Capelle was assessed as having particularly substandard internal stability. The external stability was also

Sheet pile walls were installed and fastened with tie bars to reinforce the dike

not up to standard over a stretch of 100m. The level of urgency was high but there was also no denying the impact of dike reinforcement work on the historic town. There was very little room in this fully built-up area for such alterations. The historic character and spatial quality were important factors to be taken into account. Moreover, it was a given that the local people would have to be closely involved in the preparations and execution.

Residents involved in decision making

There are various ways of reinforcing dikes; for example, you can use earth, sheet pile walls or diaphragm walls. To involve local people in finding the best solution, the regional water authority first compiled a 'most likely option' and asked local residents and businesses whether this took sufficient account of their interests and suggestions. This variant was then worked out further to produce a 'preferred option' and, in a later phase, a project plan under the Water Act. The associated environmental impact assessment showed the effects of the various alternatives. It included an examination for three dike sections, each with its own particular features and challenges, of the consequences of various solutions for local people, the surrounding area, water safety, management and maintenance. Newsletters were used to keep residents informed. They were able to air their plans, wishes and objections in round-table talks with small groups of residents, during drop-in evenings and in individual talks. In the end, a solution involving earth in combination with measures to improve the spatial quality was chosen as the preferred option (because it caused the least inconvenience and damage).

Solution using earth with high-quality design

Clay was added to the land side of the dike along almost the entire section. The slope angle was reduced and a shoulder was added to improve the stability. The modifications had to be customised to make sure they were as unobtrusive as possible and did not alter the appearance of the place. For example, the shoulder was incorporated into private gardens. The section of dike on the land side adjoining the outer moat around Slotpark had to be protected from slippage. Therefore a layer of riprap was deposited on the floor of the moat. However this was done in such a way that the moat only had to be made a little narrower, thereby preserving the historic appearance as much as possible.

Before and after: the construction of a footbridge at the Slotpark now offers a new walking route

Before and after: A customized solution in Dorpsstraat has improved spatial quality

Sheet pile walls and public gardens

The first step for the weak part of the dike on the water side was to deposit riprap at the bottom. Then a steel sheet pile wall was inserted in the crest on the water side. Holes were made in the sheet piling to prevent any disruption to groundwater flows. A special solution was conceived for a low-lying listed building on the land side of the dike: an additional stability wall in the land-side crest. The dike was made even sturdier by the use of tie rods attached to the sheet piling on either side of the dike. Thanks to this new technique, the connectors can subside in line with the subsidence of the peat soil (20 cm every 50 years) while the sheet pile walls remain solidly supported by the layer of sand below.

Opportunity to kill two birds with one stone

The project was an invasive operation but it also offered the town an opportunity to add greenery and improve public spaces. For example, Raadhuis Square obtained a better view of the Hollandsche IJssel. Decking was installed on top of the earth-retaining wall next to the outermost moat around the historic Slotpark and steps were built up the slope. This created a logical walkway connecting up Slotpark, Raadhuis Square and Doorenbosplantsoen.

Vertical gardening with flower pots in a patio garden

With the development of public space, contact with the water has been enhanced

The lowest point in the Netherlands

Capelle aan den IJssel in the province of South Holland lies within the triangle formed by Rotterdam, Zoetermeer and Schoonhoven. This area, which is managed by the regional water authority of Schieland en de Krimpenerwaard, lies several metres below sea level. An artwork in a field close to Nieuwerkerk aan den IJssel marks the lowest point in the Netherlands: 6.67 m below Amsterdam Ordnance Datum (the average sea level). In the distant past, the ground level would have been as much as 4 m above sea level but ground levels have persistently fallen as a result of peatland reclamation from the tenth century onwards, initially for agricultural purposes and later in the form of large-scale peat cutting. In cultivated land that bordered the Hollandsche IJssel and other rivers, outer embankments were built that evolved into continuous dikes from the twelfth century onwards. By the seventeenth century, the exploitation of peatlands had turned the area into a patchwork of pools and lakes. Windmills were built all over the countryside to drain them again. This is now one of the most densely populated parts of the Netherlands.

The Zuidplaspolder is the deepest polder in the Netherlands

05
08
Sample Sheet of Intervention – overview on page 24-27

A construction kit and 200 volunteers

Programme: **Major Rivers Delta Programme**
Province: **Overijssel**
Coordinates: **52°33'31.5"N 5°55'03.6"E**
GPS: **52.558762, 5.917664**
Assignment: **to protect the historic town centre from flooding**
Implementation (detailed plan through to construction): **1997-2002**
Water board: **Drents Overijsselse Delta water board (previously Groot Salland water board)**
Design engineer(s): **BDG Architects, in collaboration with the consultancy firms Witteveen + Bos and Royal Haskoning**
Contractor/consortium: **Koninklijke Woudenberg in collaboration with Van den Herik and Van den Biggelaar BV**
Stakeholders: **the municipality, local residents, local businesses, historic buildings department (only for the historic buildings in the flood defence)**
Project area (ha): **the flood defence extends for almost 2 km**

The town of Kampen has a historic centre adjoining the River IJssel. The 2 km long IJssel embankment constitutes the river front. The town is exposed to threats of flooding from two sides: high water levels in the IJssel and the water in the IJssel Lake being pushed up by a severe north-westerly storm. Things can get particularly bad if both happen at once. In 1995, the embankment was submerged in water and the town centre was only saved from flooding by an emergency dike and countless sandbags. The entire dike ring from Kampen to Elburg was raised in height as part of the flood protection programme but Kampen town centre presented a problem. Raising the embankment would upset the appearance of the town, which has listed status, and disturb the town's relationship with the river. The solution turned out to be a demountable flood defence that can be assembled quickly and effectively by a large group of volunteers. This Flood Brigade has about 200 members, subdivided into teams each of which is responsible for a specific part of the structure.

The town as a flood barrier

The demountable structure is a construction kit intended to seal off the town centre's riverfront and protect it from flooding. The barrier is formed by the existing medieval town wall, which has largely

The passage in the IJsselkade is closed during high water

Lifting gates such as this one in the Melkpoort are inconspicuously recessed when not in use

The IJssel near Kampen, engraving c. 1870

DEL PORTO

Restaurant

The built-up flood defence is inspected by bike. Behind the bikes a map showing the locations of all openings that have yet to be closed

The water authority's storage facility where all the necessary materials are stored in an orderly fashion

been incorporated in the buildings on the IJssel embankment. All passageways, openings and mouse holes have to be closed off. This involves a total of 84 openings, such as alleyways and squares, and 220 flood protection aids placed in front of doors and windows. Each element in the barrier requires its own tailored solution, and all the parts have been numbered and stored carefully. Traditional stop logs are used in many places; these are planks that are stacked up between rabbets to form a flood protection barrier. These days they are made of aluminium and have rubber profiles, making them watertight when fitted together. Lift-and-slide barriers and flap barriers are used in the alleyways and narrow streets. Lift-and-slide barriers are sunk into the ground and have to be hoisted up. Flap barriers lie flat on the street and only have to be pulled up. Other solutions are flood protection doors, sluice doors, glass sheets and flood defences in the ground. To make sure the town really could be made watertight, pipes and sewers had to be rerouted and the soil remediated in places. The work presented a good opportunity to restore the historic Oude Buitenhaven harbour, redesign the embankments and remediate the contaminated soil.

Flood Brigade made up of volunteers

The water defence system can only work if the town's inhabitants cooperate. No fewer than 233 properties are part of the barrier. Work has to be carried out in 32 houses and 52 gardens to erect the barrier. In six cases, the barrier runs right through a home and watertight doors have been installed inside. The barrier also crosses a garden and three business premises. The water board has keys to all the front doors so that it can gain access where necessary in the event of an emergency. A second form of cooperation is the volunteer Flood Brigade that erects the barrier. If there is a serious threat, the Flood Brigade assembles in the depot in IJsselmuiden, on the other side of the river. Like in a fire station, this has containers where the barrier parts are permanently available for use. Once a year, the brigade practices putting up the barrier. Each team knows what to do and can look this up in the manuals. All the elements are checked immediately once they have been erected and the water board maintains tight control over everything. In this way, it is possible to set up a complex flood protection barrier very quickly using what are essentially very simple elements. The cooperation between professionals and volunteers in the guise of the Flood Brigade makes Kampen's water defence system unique in the world.

Installing the flood defence is rehearsed annually

The stop logs are transported in waste containers after which the team of volunteers can quickly install the flood defence

This building is part of the flood defence too

02
03
09
Sample Sheet of Intervention – overview on page 24-27

Spakenburg: a buoyant barrier for a village with protected status

Programme: **Second Flood Protection Programme**
Province: **Utrecht, Gelderland**
Coordinates: **52°15'14.9"N 5°22'41.1"E**
GPS: **52.254127, 5.378082**
Assignment: **water safety along the River Eem and Eem Lake**
Implementation (detailed plan through to construction): **July 2012 - 2017**
Water board: **Vallei and Veluwe water board**

Subproject in Spakenburg town centre:
Design engineer(s): **Grontmij**
Design of floating barrier: **Johann van den Noort**
Contractor/consortium: **Van Heteren (Hengelo)/ Jansen Venneboer (Wijhe), Hyflo Self Closing Flood Systems BV, Aggéres (emergency barrier technology)**
Stakeholders: **advisory group with representatives from Natuurmonumenten (society for the preservation of nature in the Netherlands), the Nature and Environmental Federations of Utrecht and Gelderland, the Eemvallei Preservation Foundation, the Grebbelinie Foundation, RGV Holding BV, Pos sportvisserij, the Agricultural Businesses sounding board, North branch of Dutch Federation of Agriculture and Horticulture, Gooi en Omstreken Birds Workgroup, 'Aan de dijk' Society, WSV De Eendracht, residents of Oostmaat and surrounding area, and the civil engineering firms Witteveen+Bos and Grontmij**
Length: **335 m**

Spakenburg is an old fishing town that used to lie on the Zuiderzee Bay. In the infamous storm surge of 1916, when dikes around the Zuiderzee Bay were breached and there was major flooding, Spakenburg was one of the places that suffered severely. The construction of the Afsluitdijk causeway in 1933 changed the bay into a lake, the salt water was replaced by freshwater and new land – the Flevopolder – was reclaimed a few hundred metres off the coast. Yet flooding remained a threat whenever a severe north-westerly storm occurred in combination with high levels of discharge in the River Eem. To meet the safety standards, 22 km of the 45 km of dike along the southern border lakes and the Eem would have to be raised in height, reinforced or protected against penetration by seepage water (known as 'piping'). This was a huge operation, which was subdivided into the following projects: Jericho, Grebbeliniedijk, Eemdijk-Zuid, Eemlandse en Slaagse Dijk, Kern Eemdijk-Zuid, Westdijk Eemdijk-Noord, Kern Spakenburg, Oostmaat, Zeedijken and engineering structures. Dedicated solutions were conceived for a number of places. For example, the new residential districts of Jericho and Jerusalem in Amersfoort are encircled by a wall rather than a dike. This wall is 500 m long, 80 cm high and incorporates a strip of greenery. The biggest challenge,

The old port in Spakenburg where the water can surge to high levels

though, was presented by the two areas of historical significance: the town of Spakenburg and the Grebbe Line dike, both of which have protected status. The task was to improve water safety without damaging their cultural and historical qualities.

Spakenburg: town with historic character

Spakenburg's old harbour has protected status as a small town centre with historic character, and it is a significant tourist attraction. Raising the embankment or closing off the harbour was out of the question, but somehow or other it would have to be possible to increase the height of the embankment over a length of more than 300 m. The solution was conceived 75 km away during the infamous high water levels of 1995. The artist and garage owner Johann van den Noort in Kampen had been spending all day moving his things into safety from his business premises just in case the dikes were breached. He was amazed that they were having to build an emergency dike made of sandbags to protect Kampen town centre. The story goes that it took him only 15 minutes that night to come up with the idea for an emergency barrier that is put in place by the swelling waters, as it were. The principle is simple. When water levels rise, the water flows into a steel tank under the embankment via inlet openings and tubes. The tank houses a synthetic (kevlar) water barrier that floats upwards as the water flows into the tank. The water pressure ensures that the barrier stands proudly upright for as long as water levels are high.

Practical applications

It took years before the idea had been worked out in detail and Van den Noort had gained a foot in the door of the flood protection sector. His first floating barriers were built in other countries. The Van Abbemuseum in Eindhoven, which is right next to the River Dommel, also uses a barrier of this type to protects its storage rooms. Kampen opted for a conventional water defence system that is put in place by volunteers. Spakenburg is the first place to use this innovative solution as the primary flood defence , with a barrier 80 cm high and 335 m long. The work that was carried out in the town consisted of reinforcing and partially rebuilding and raising the existing embankments. The self-rising emergency barrier was installed under the embankment. Normally only the top of the barrier is visible as a strip of COR-TEN steel in the road surface. The inlet openings are fitted in the embankment wall. Once the water starts to flow through these openings, the emergency barrier is pushed upwards. There is a safety switch to prevent parked cars or other objects on the embankment being inadvertently jacked up. That means human intervention is still required to turn the sign to green for the barrier to operate. The Spakenburg town anthem has been engraved in the barrier's cover.

Grebbe Line dike: historic landscape

In addition to the former Zuiderzee dike, the dike along the River Eem was reinforced too. The Grebbe Line, which was built in the eighteenth century and connected the Lower Rhine river at Rhenen to the Zuiderzee bay at Spakenburg, passes through

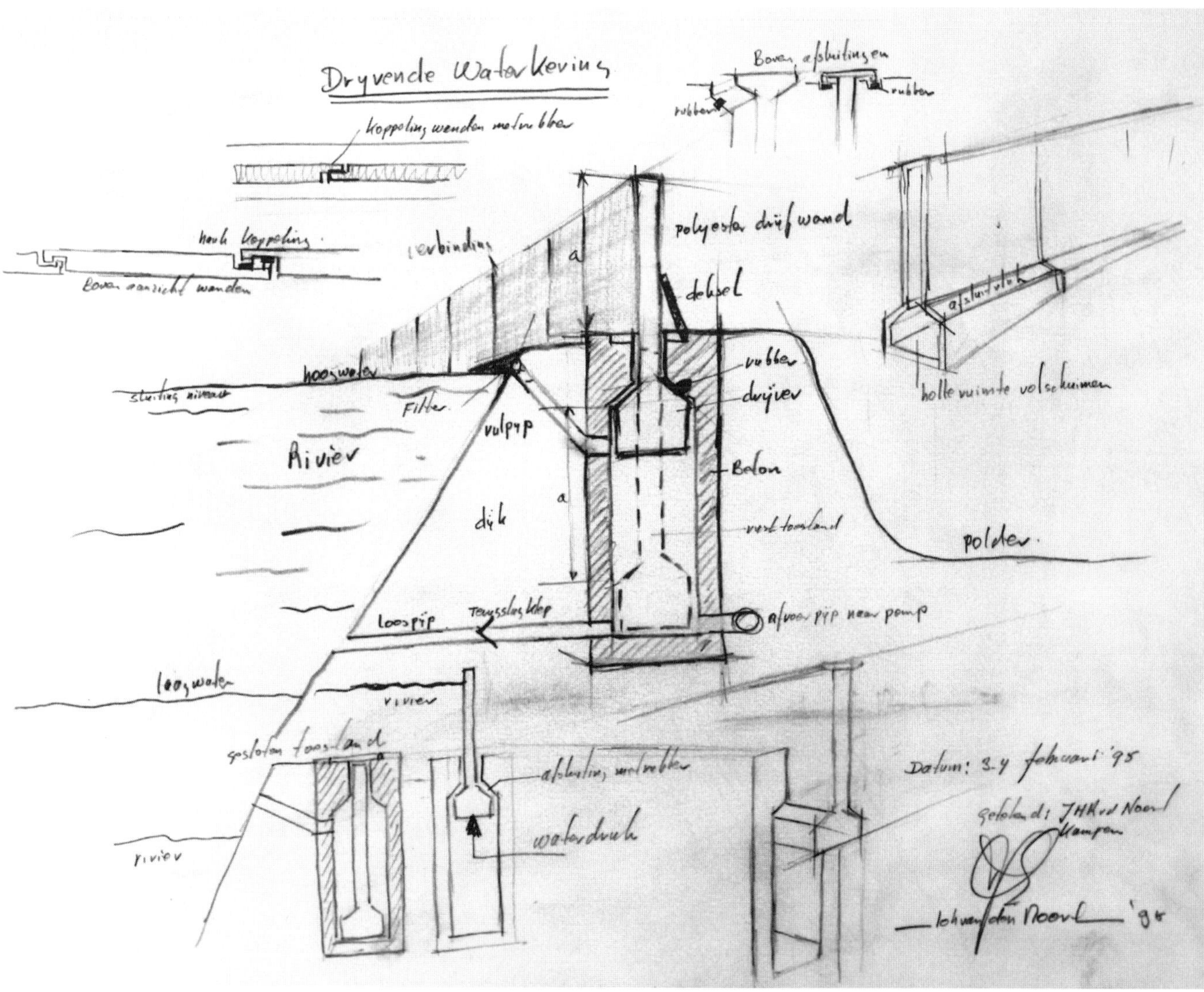

The design sketch for the flood defence that Johann van Noort made

here. There was a dike along the Eem, embankments and fortifications, along with barrages and sluices that could be used to deliberately flood the surrounding area in a controlled fashion in the event of a war. The line was operational until after the Second World War and it was regularly upgraded and strengthened. The Grebbe Line was an outpost in the defence of the big cities in the west of the Netherlands; its purpose was to hold up or repel an advancing enemy army. That would give the Dutch time to get the Nieuwe Hollandse Waterlinie (see p. 78) near Utrecht in a state of readiness.

Integrating into the landscape

These days, the Grebbe Line is a listed monument. Prior to the dike reinforcement work, an extensive archaeological survey was conducted and the distinctive features of the landscape, the green structure and the buildings of cultural and historical interest were mapped. The dike reinforcements were designed to avoid any impact on the archaeology, listed buildings, protected trees and unusual structures. The design was geared to the surroundings to preserve the landscape's appearance and cohesion. An additional complication was the fact that the dike runs through 80 private gardens in the village of Eemdijk. Preparations for the reinforcements were made in collaboration with the residents and took years. The work itself could be completed quickly. After obstacles and plants had been removed from the dike, small-scale machinery was used to add a layer of clay to raise the dike to the required height. Then the gardens were laid out once again.

The flood defence lies hidden in the quay. The Spakenburg anthem has been engraved into the covering of the flood defence

The flood defence when it is raised

The Eem starts in Amersfoort and flows into the Eemmeer

The Eemdijk runs through 80 private gardens, 2014

International Aspects

Water is a priority sector for the Netherlands. We aim to lead the way internationally in water technology, maritime technology and delta technology. Ever since the days of the Dutch windmills, other countries have looked in amazement and admiration at how the Dutch have tackled water management and the development of the delta. All over the world, cities and areas can be found that were built by the Dutch or inspired by the Netherlands' hydraulic engineering. The story goes that the French philosopher Descartes once said 'God created the Earth, but the Dutch created Holland'. It seems almost like a fairy tale, one that has yet to end. The *New York Times* recently ran the headline: 'The Dutch Have Solutions to Rising Seas. The World Is Watching'. The newspaper describes how the Dutch live with the water in the rivers and sea rather than fighting it. The paper concludes that climate change is a business in the Netherlands, like cheese in France or cars in Germany.

In Both Directions

Yet the Netherlands is not an oasis in a desert full of climate sceptics. Hydraulic engineering has always taken inspiration and input from contacts with other countries. For example, in March 1953 a young Argentinian engineer landed at Schiphol, eager to learn more about Dutch hydraulic engineering. The North Sea Flood had hit Zeeland only one month before so José Loschacoff (born 1927) could set to work immediately. He never left Zeeland and became a director in the government Delta department. Loschacoff cut quite a dash amidst the Dutch engineers who were good at mathematics but rarely read a book and could not dance at all. He brought his cultural background with him and involved landscape architects from Staatsbosbeheer in his designs. At the drawing table, dams were transformed into complete landscapes full of new potential uses. Loschacoff: 'Bringing in architects, especially landscape architects, hardly adds to the cost but gets so much better results.' That is precisely how the Netherlands is able to make a difference – by linking safety to quality.

A Good Idea Draws on Many Sources

Nicolau Klüppel (1930-2016), nicknamed Nicolago ('Nico-Lake'), was a Brazilian engineer and contemporary of Loschacoff. He was the water expert in the team of the architect and mayor Jaime Lerner, who transformed the city of Curitiba into an ecological metropolis. In the 1990s, hundreds of urban planners and politicians from all over the world travelled to Brazil to see what had been achieved in public transport, urban planning, culture, education, waste processing and city marketing. In 1966, Klüppel came up with the idea of laying out linear parks along the river banks. This communal function prevented slums from developing. Nicolago said no one would be so stupid as to build a house on the beach when the tide is out, but if a riverbed floods once every ten years, an entire town has been built there before you know it. Because the Curitibans yearned for the sea and water, secondary channels and lakes were dug in the parks from 1971 onwards. This solved the problem of flooding while improving spatial quality. Nature conservation, recreation and water safety went hand in hand. Years later, the Netherlands would adopt the same course of integrated area development. The experience of Curitiba shows that good ideas can emerge independently in different places and that the river widening approach is not only successful in rich countries. The ideas obtain their value from the way in which they are implemented, acquire public backing and help tackle the local challenges.

Cross-Border

Back to the Netherlands. However large the scale on which the water issue is tackled in this country, the problem of the delta remains a cross-border problem, covering the river basin of the major rivers across Europe. The Grensmaas project deals with the landscape in the Netherlands and Belgium. The river widening project at Meers involves work on the RivierPark Maasvallei, a Dutch-Belgian project. The Rhine delta extends into Germany. In 2014, there was a great risk of flooding in the German Lower Rhine due to the poor state of the German dikes. This constituted a threat for Gelderland. Centuries ago, it had been Gelderland that had been failing to keep up the maintenance of its dikes. The Dutch in the province of Holland built the Diefdijk to put an end to the flooding caused by their neighbour. Today, a border dike would be out of the question. The solution now lies in cooperation. The Dutch student Robbert Jongerius showed what this can achieve in his graduation project on the German Lower Rhine. He drew climate-proof dikes forming a recreational link between Germany and the Netherlands. That would give an additional 15,000 ha in retention capacity that could be designated for use as a national park. Riparian woodlands would appear in the new channels, linking the nature along the Lower Rhine to the drier wooded areas along the edge of the valley. Hydro-electric turbines would supply energy. The project shows the potential for forging new links and enriching the area through water safety measures.

Exporting expertise on deltas

In 2015, Henk Ovink was appointed the Netherlands' first Special Envoy for International Water Affairs. In this capacity, he travels all over the world to put water safety and climate change on the agenda, encourage innovative solutions for climate change and, when requested, provide expertise on specific questions. He has two key messages. First, that solutions to the complex problems are found through collaboration. Second, that water safety and quality always go together. The mindset is what counts. This creates opportunities for improvements, public awareness and an attractive business climate, which is needed to make towns and cities resilient and resistant to the impact of climate change. Only rich countries such as the Netherlands can afford to spend tens of millions of euros on a high water channel that may fill with water only once every eighty years. But other countries can benefit from the integrated approach to water safety as practiced in the Netherlands. This enables specific solutions within the scope of the possibilities and limitations of that location. Ovink:

The new Lake Borgne Surge Barrier will protect New Orleans and its environs against flooding

The West Closure Complex Pump Station near New Orleans can quickly drain large quantities of rainwater to reduce flooding

'The Netherlands undeniably leads the field in water governance, water management and designs for our water-rich country. No other country has made water such a key part of the area development culture, in which continually learning and innovating helps time and time again to produce a better, safer country. Dutch experts work all over the world implementing this integrated approach, however difficult that may be in many places.'

Bangladesh, Peru and Mexico

The Ganges and Brahmaputra delta in Bangladesh is somewhat reminiscent of the Netherlands, however different both countries are. Dutch consultants advised the country on the construction of shelters and escape routes in the event of flooding. And that paid off. The next time the country was hit by a storm, the number of fatalities was in the hundreds rather than thousands. Countries such as Peru and Mexico are drawing inspiration from the Dutch by taking an integrated approach to their water problems. A great deal can be achieved by giving safety and quality top priority. There are no fast, one-dimensional solutions for tackling the water challenge. Water management is a way of life.

New Orleans and New York

After the devastation caused by Hurricane Katrina (2005), New Orleans has worked hard on its recovery. When the city was hit by the next hurricane, it was spared flooding thanks to the dams and storm surge barriers, which were built in part thanks to Dutch expertise. The next challenge is to embed climate resilience in the city's culture: in management and maintenance, public administration, the link with ecological and economic issues, and in the city's qualities and potential. The Greater New Orleans Water Plan is the result of a Dutch-American partnership in which an integrated approach was taken focusing on the city's opportunities. New York was badly affected by Hurricane Sandy in 2012. Henk Ovink was asked to help with the reconstruction of the region. He did so by organising the Rebuild by Design Competition, followed by the National Disaster Resilience Competition. In both cases, strong contributions were made by both local communities and the Netherlands. The Netherlands has a real edge over the rest of the world thanks to the centuries of experience with water control systems, but it will need to keep refreshing its expertise and translate its experience into solutions for present-day challenges.

Michael Kimmelman, *The Dutch have solutions to rising seas. The world is watching.* NYT, June 15, 2017

Many inhabitants of Bangladesh live near the water, which makes them vulnerable to flooding

Dutch firms Arcadis and ONE Architecture participated in the design of the BIG U, which must protect Manhattan against future storms

Urban Parks and Recreational Areas

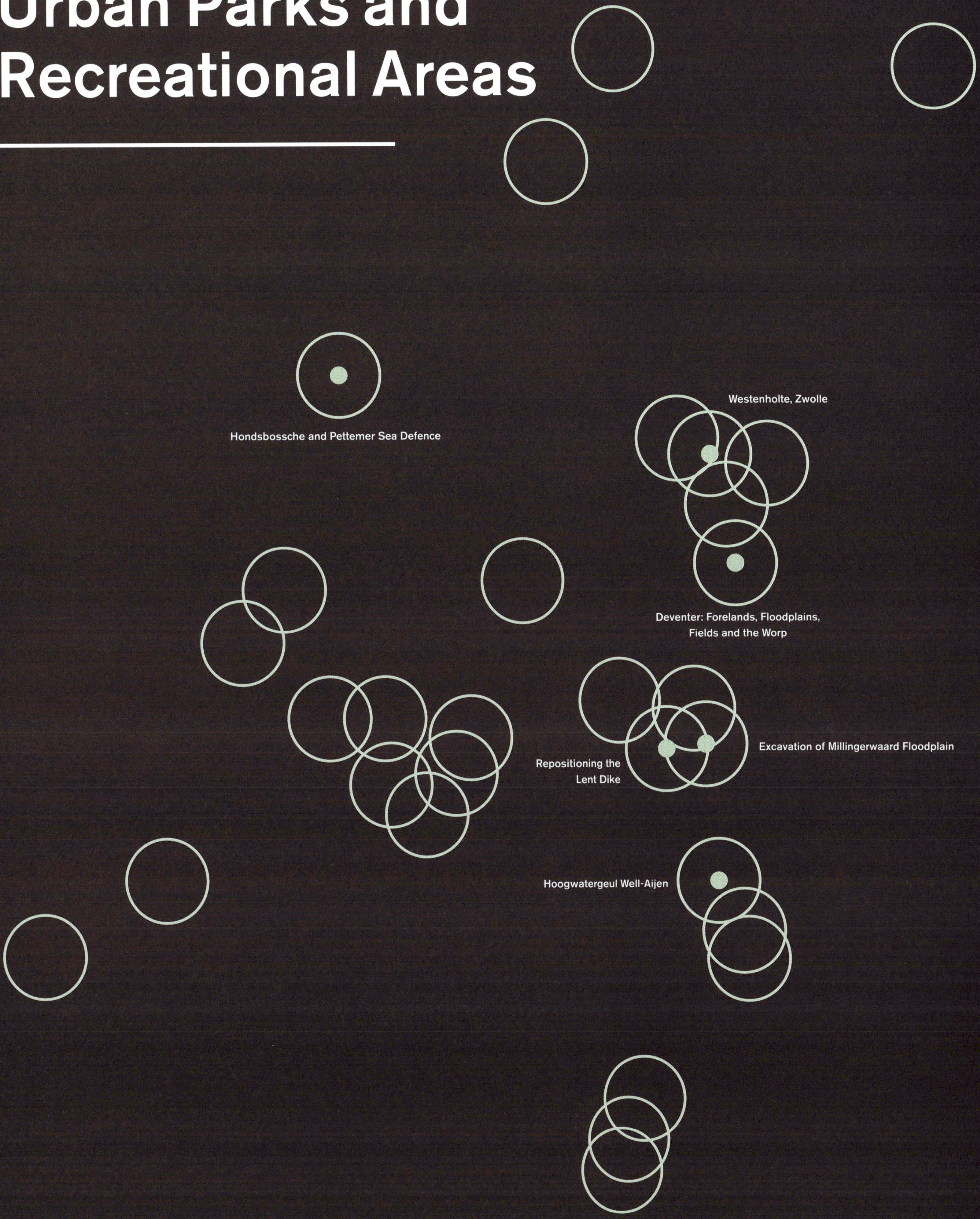

Repositioning the Lent Dike – The city island near Nijmegen provides wonderful recreational opportunities, 2016

Hondsbossche and Pettemer Sea Defence – A 25-meter-high climbing dune has been built at Petten, 2015

I DON'T
WANT
TO LOVE
YOU
BUT I
LOVE
YOU

Sample Sheet of Intervention – overview on page 24-27

Beach and dunes take over from retired dike

Programme: **Second Flood Protection Programme/Reinforcing our Coast (Kust op Kracht)**
Province: **North Holland**
Coordinates: **52°44'23.3"N 4°38'44.1"E**
GPS: **52.739802, 4.645595**
Assignment: **dike reinforcement**
Implementation: **2014-2015**
Water board: **Hollands Noorderkwartier regional water authority**
Design: **Arcadis**
Contractor: **Boskalis and Van Oord**
Project Manager: **Directorate-General for Public Works and Water Management (maintenance of beach and shoreface), Holland Noorderkwartier regional water authority (primary sea defence)**
Length: **400 m on the land side of the dike and 100 m on the sea side**
Earthworks: **35 million m^3 of sand**
Other figures: **14 ha of marram grass planted**

For centuries, the coastal area of North Holland was protected from the North Sea by a continuous chain of dunes. However, the dunes were battered by severe storm surges, which severely damaged the dunes between Camperduin and Petten time and time again. Even by the Middle Ages, these dunes had already almost completely disappeared. They could be restored with the help of sand, pileworks and longitudinal groynes but this solution never lasted long. In 1872 the dike-reeve Cornelis van Foreest had the revolutionary idea of replacing the marram grass growing on the sea side of the sand dike by hard basalt blocks. The waves could run onto the blocks without causing any damage.

In 1877, work finished on the Hondsbossche Sea Defence – named after the wood (*bos*) that covered an area equal to one *hond*, an old Dutch unit of measurement, before it was destroyed by the All Saints Flood of 1570. The Hondsbossche Sea Defence was a great success. This dike structure has only recently lost its function as a sea defence. It is now enjoying a well-earned retirement as a monument to the struggle to protect the country from the sea. The hinterland is once again protected by a continuous chain of dunes. One spin-off from the new sea defence is the extension of the coast with a beach and dune area suitable for recreational purposes.

The old Hondsbossche Sea Defence at Camperduin served faithfully for 130 years, 2010

21

23

Halfway through the metamorphosis, 2014

Camperduin has been given a new dune area, 2017

Rising sea levels

What had happened? The continuous erosion of the adjoining dunes meant that the Hondsbossche Sea Defence, which forms a 5.5 km barrier together with the Pettemer sea defence, was increasingly far out to sea. The dike had been raised to the Delta-Plan height (of 11.5 m above Amsterdam Ordnance Datum) back in 1981. That seemed to be sufficient until new information on increasing sea levels became available in 2003. The force of the waves crashing onto the coast in heavy storms would far exceed the force that the dike had been designed to withstand. The Directorate-General for Public Works and Water Management and the Hollands Noorderkwartier regional water authority immediately took some emergency measures while awaiting a more permanent solution. They placed a temporary 70 cm high steel sheet pile wall on the Pettemer sea barrier and they fitted the Hondsbossche Sea Defence with stronger, rougher cladding.

Building with the help of nature

In the search for a sustainable alternative, 'hard' solutions such as raising the crest of the dike or building an overtopping-resistant dike were soon rejected. The height of a dike cannot be increased ad infinitum as this makes them vulnerable. And in this instance the dike would have to be raised by as much as 4 m. Therefore the decision was made to adopt the principle of building with the help of nature. A total of 35 million m^3 of sand was used to extend the coastal defence on the sea side of the dike by adding a series of dunes, a 50 m wide beach and a shoreface. The protective dunes running the entire length of the dike are 75 to 100 m wide and are 7 to 12 m above Amsterdam Ordnance Datum. One steep dune near Petten even reaches a height of 25 m. The protective dune has been built right next to the sea-side shoulder of the old dike. The coastline has shifted 300 m out to sea.

Dynamic and flexible

A solution using sand is a dynamic solution. Any sand that is blown off the dunes in stormy weather is deposited on the beach and the shallow underwater bank. When the weather is calm, the movement of the water and the wind pushes the sand inland and the dunes grow again. The marram grass growing on the dunes also has an important task. The grass prevents the sand from blowing away during storms and secures the sand that is deposited on the dunes by the wind during calm weather. Sand dunes can recover of their own accord and grow in step with rising sea levels. They can easily be extended and modified and they keep

In Petten a dune is constructed in front of the old sea dike, 2015

the sea at a safe distance from built-up areas. What is more, this natural solution did not take long to construct: the required level of protection had been achieved nine months before the deadline – the end of 2015. The contractor will remain responsible for maintenance for the next 20 years.

Local area contributes ideas on how to improve the spatial quality

In a project entitled *Kust op Kracht* (Reinforcing our Coast), the Directorate-General for Public Works and Water Management and Hollands Noorderkwartier regional water authority are also tackling other weak links along the North Sea coast, including at Callantsoog and Sint Maartenszee. An important task for the entire project was to combine measures to strengthen coastal protection with improvements in spatial quality wherever possible. The province of North Holland, the municipalities of Schagen and Bergen, Natuurmonumenten and Landschap Noord-Holland worked jointly on implementing this task. They gave local residents and business people the opportunity in information meetings and design studios to contribute ideas on the development of nature and recreation in the coastal area. This input was passed on to the contractors taking part in the tender procedure. This interactive approach allowed good ideas and opportunities to be incorporated in the final design for the coastal defence. There is broad support for the solution because everyone can see their own wishes reflected.

Attractive for humans, flora and fauna

It is now possible to walk along the high-water mark in the new dune area, which was rechristened the Hondsbossche Dunes after a competition, and cycle along the protective dune. A lagoon was created near Camperduin while Petten has acquired a panorama dune and a 50 m wide beach. The results of a nature survey were announced in early 2017. The operation had only a limited negative impact. While fewer oyster catchers, waders and whimbrels were counted, the nature area of De Petten with its little islands was used as a breeding ground in 2016 for the first time by at least 7,000 Sandwich terns. Another surprising finding was that rare plants such as samphire, sea spurge, sea fern-grass and frosted orache had managed to establish themselves on the old dike, which is now covered with sand.

The new dune in Petten, stillunder construction, half a year later, 2015

More than 14 ha of marram grass was planted to prevent sand drifts, 2015

The new dunes, with the climbing dune in the background, 2015

Sample Sheet of Intervention – overview on page 24-27

14
15

Dreaming of a large area for nature and recreational areas

Programme: **Zandmaas Meuse Works**
Province: **Limburg**
Coordinates: **51°33'59.8"N 6°03'02.6"E**
GPS: **51.566614, 6.050713**
Assignment: **reduce the high water level**
Implementation: **2010-2014**
Water board: **Limburg water board (formerly the Peel en Maasvallei water board)**
Design: **Directorate-General for Public Works and Water Management, Meuse Works**
Contractor: **Consortium Kampergeul BV**
Project Manager: **The Limburgs Landschap Foundation**
Drop in water level: **7 cm (26 cm in 2025)**
Area: **19.4 ha**
Earthworks: **286,000 m³**
Other figures: **127 + 40 ha of new nature, 3.5 km of ring dike**

Ever since the early 1990s, the northern Limburg municipality of Bergen had dreamed of turning the land between the Meuse, the nature area of Maasduinen and the villages of Aijen and Well into the largest area dedicated to nature and water-based recreation for miles around. Who would have thought back then that this dream would be fulfilled thanks to major safety measures? The construction of two high water channels one of the measures decided upon in 2002 in the Tracébesluit Zandmaas/Maasroute (Zandmaas/Meuse Route transport infrastructure decree) created an ideal opportunity for integrated area development. What is more, the costs of the 600 ha Maaspark Well nature and recreation area, which is scheduled to be completed in 2025, can be covered almost entirely by the proceeds from the 16 million m³ of sand, earth and other raw materials excavated during the construction work.

Secondary channels

The northern high water channel near Aijen and the southern one near Well, which is a continuation of the Aijen channel, are directly connected to the River Meuse. They form inlets that flow with the Meuse when water levels rise, thus ensuring the rapid, safe discharge of excess water. When water levels are low, the water remains still. They are both laid out with a number of promontories and they vary in depth from 2.5 to 7.5 m. The banks can be gently sloping or steep and parts can dry out while other parts remain marshy.

Risk management

The northern channel was scheduled to be tackled first. The required reduction

A fishing spot at the Leukermeer

in water levels of 7 cm by the end of 2015 was in danger of not being achieved due to delays. It was possible to achieve this interim target on time after all thanks to the expeditious decision to bring the work on the southern channel forward. The following target – 11 cm in 2016 – was also met. Ultimately, a reduction in high water levels of 26 cm has to be achieved by around 2025. The protective effect when water levels are high will be noticeable up to tens of kilometres further upstream. The construction of the southern channel even offers possibilities for further river-widening measures in the future. The work has had a beneficial side effect in limiting local erosion of the banks and sedimentation downstream in the Meuse. That saves on maintenance costs.

Paradise for fans of water sports

The municipality of Bergen is taking various measures to develop the area for recreation. For example, the current Leukermeer lake outer harbour is being expanded and equipped with 400 berths. A striking watergate is being built at the entrance to the Meuse. Rowers and canoers can use the channel. Distinctive recreational facilities with accommodation are being developed on the water's edge. A water boulevard connects Leukermeer lake to the Maasduinen National Park visitors' centre, which opened in 2012. Maasduinen has a very different landscape to that of the river valley but it too was created by the river. The westerly winds have produced high, parabolic-shaped river dunes on the east bank of the River Meuse. This area extends from Venlo to Gennep. New cycle paths and hiking trails are giving people access to the river valley and river dunes.

Nature managers, farmers and beavers

The Maaspark Well area covers 127 ha of new nature. Setting up the high water channel as a nature area will be of significant benefit to the ecological performance of the River Meuse and its floodplains. A wide variety of flora and fauna will be able to thrive in the drier and wetter spots. Limburgs Landschap is responsible for the nature management. It is looking to collaborate with local farmers. Beavers function as a natural form of bank maintenance as they keep tree growth under control and create open spaces. The dams that they build make the water flows more dynamic. That benefits aquatic life, dragonflies and frogs.

Archaeological traces

The slightly elevated spots in the river valley between Aijen and Well where sediment has been deposited have been the site of settlements for centuries. Preliminary archaeological surveys found remains of homes from the New Stone Age, the Iron Age and the Roman era under the thick layer of clay that has been deposited during floods over the past 2,000 years.

One end of a newly created southern secondary channel at Well

All sorts of recreational activities take place along the shores of the Leukermeer

New secondary channel at Well with two branches

Compensation area for badgers

Old river dunes along the Meuse are an ideal habitat for badgers. Underground setts can be found all over the place. These are colossal structures with an extensive network of passageways and chambers, sometimes with several storeys. This rare, protected animal feels at home in the project area too. Excavations were designed to minimise the impact on the badgers but a number of setts had to be destroyed even so. Because of this, a compensation area covering nearly 9 ha is being established to the south of Aijen. Given that the badgers are able to move into two abandoned setts, it was decided to make this a foraging area. It is being laid out in such a way to make it as accessible as possible, with plenty of food and sufficient shelter. The nutritious soil is a source of earthworms. The grazing cattle will keep the grass short enough for the badgers to find the worms. What is more, the badgers can vary their diet by feasting on the fruit trees in the orchard or amphibians in the surface water. They can move unseen along hedge banks, hedgerows and ditches. The steep banks of the ditches are also ideal for building new setts. The Das&Boom ('badger&tree') foundation is closely involved with the layout plans and monitoring. Once the surrounding area has returned to a state of peace and quiet, it will hopefully not take long before the new lodgers discover the foraging area.

A badger 'compensation' area has been created on the north side

The compensation area will develop as a river dune

Sample Sheet of Intervention – overview on page 24-27

Flood safety thanks to faster nature development

Programme: **Further Elaboration of Rivers Area (NURG)/'Room for the River'**
Province: **Gelderland**
Coordinates: **51°52'20.1"N 6°01'28.2"E**
GPS: **51.872250, 6.024500**
Assignment: **allocating water to the Pannerdensch Canal and the River Waal, increasing capacity**
Implementation: **2013-2020 (project decision in 2012)**
Water board: **Rivierenland Water Board**
Design: **Staatsbosbeheer in collaboration with the province of Gelderland, the Directorate-General for Public Works and Water Management, the municipality of Berg en Dal, Rivierenland water board, the ARK foundation, and residents and stakeholders in the area**
Contractor: **consortium of Boskalis/Van de Wetering in partnership with K3Delta (earthworks), contractor Den Boer BV (demolition and remediation)**
Project Manager: **Staatsbosbeheer in partnership with the ARK Foundation**
Drop in water level: **9 cm**
Area: **420 ha**

When the Rhine enters the Netherlands at Lobith, it splits into the Lower Rhine and the Waal (RR10). Two thirds of the water enters the Waal, which meanders towards Nijmegen in three large U-bends. The Millingerwaard, an extensive floodplain, lies on the first of the inside bends. When water levels rise in the Waal, the water washes over the Millingse Dam, just before the bend, and into the floodplain. The Waal does not flow so fast in the bend. Sand shifting and deposits by the river have created a dynamic dune area here. At a height of 10 m, the Millingerduin is the highest river dune peak in the Netherlands. The alluvial ridge is partly wooded. These so-called riparian woodlands are on marshy ground. The Colenbrandersbos wood is a rare remnant of a hardwood riparian woodland with oaks, elms and hawthorns. There are softwood riparian woodlands in the boggy areas with black poplar trees and white willows. The aerial roots are reminiscent of mangrove woods in the tropics.

Precursor for nature development

The transformation of Millingerwaard began years before the start of the water safety projects. In 1991, the ARK foundation started an experiment with nature development, which was extended to cover the entire area after the World Wildlife Fund and the nature management agency Staatsbosbeheer became involved

14

15

Softwood riparian woodland with white willows

(in 1992 and 1993 respectively). The project had a huge impact: fertile clay soil land was taken out of use as farmland to became a nature area outside the dykes. Even then, exotic animals were being used to maintain the terrain, such as Polish primitive horses, Galloway cattle and beavers that had been reintroduced into the Netherlands. The results of this restoration of nature were spectacular from the start.

Water safety linked to nature development

The area traversed by the main rivers between Arnhem and Nijmegen is crucial to water safety in the Netherlands. This is where the water divides into the Rhine, IJssel and Waal and where this process can be adjusted. What is more, the upper reaches of the delta are the ideal spot for creating a buffer when water levels rise: everything that is kept back here relieves the rest of the country. The assignment for Millingerwaard was to improve the allocation of the water flowing in to the Waal and Lower Rhine and to reduce water levels in the Waal by 9 cm. The solution followed on from the nature development that had already been set in motion, by modelling the natural processes in the floodplains in such a way that the water safety target could be achieved.

Konik horses next to petrified oak tree trunks

In recent years, the number of beavers has been growing in the water-rich Netherlands

In the 'seepage fingers' the clear seepage water from the floodplains bubbles up to the surface

Digging back in time

The excavation of Millingerwaard is an excellent example of making work of work. The project has multiple objectives: developing nature (265 ha), excavating raw materials (up to 2020), providing recreational facilities and, let it not be forgotten, ensuring water safety. In essence, the intervention involves reversing, as it were, the natural development of the floodplains by digging up the clay that has been deposited in the course of many centuries. This 'contour-guided clay extraction' is an interesting option from a nature development perspective as it uncovers the old pattern of channels and sand ridges. Many different rare species thrive in this poor sandy soil.

Seepage fingers with pure water

The second step consists of excavating a 3 km long channel in the flow direction and five 'seepage fingers'. The main channel serves as the most important subsidiary route for the River Waal, reducing water levels in the Waal by 6 cm. In the seepage fingers, clear seepage bubbles up to the surface from the floodplains, creating a biotope with unusual aquatic plants, insects and other species. When water levels rise, the fingers are connected up to the river. As soon as water levels fall, the water in the floodplains subsides and is discharged into the Waal via the main channel. A barrier has been built, though, to make sure the nature area stays boggy.

Trade in sand

Millingerwaard will have a few deep sand pits until 2020. Like the clay that has been excavated, the sand too can be sold in the same way as the gravel from the Meuse Works (see p. 156, 164). The proceeds can be used to finance part of the redevelopment. The pits will be filled with material that

Sand extraction in the Millingerwaard, 2015

Galloway cattle and day-trippers look for shade

has been excavated from the channels and is unsaleable; that material will therefore not have to be removed from the site. The pace of the entire operation will be determined by the gradual extraction of the clay and sand. This process will continue until 2020, at which point this industrial activity will stop. The area will then be dedicated entirely to nature and recreation. Traces of the companies will be preserved after they have gone as relics of the modern era, namely a shelter belonging to a brick factory and a chimney. The road to the gravel and storage business will become a hiking trail and cycle path after the business has moved. Fossilised tree trunks from prehistory that were found during dredging operations in the Waal in 1995 have been placed as tree monuments alongside the river. Here too a 'Woodhenge' has been created as in Meers in Limburg (see p. 156).

Gelderse Poort

Millingerwaard is part of the Gelderse Poort, an area covering 6,105 ha that is managed by Staatsbosbeheer. The entire area is open to the public: nature conservation and nature experience go hand in hand. The landscape is neither a reconstruction of the past nor a newly designed area. Natural processes (flooding, erosion, shifting sands and seepage) are being used to give the river areas back its old ecological values in a new form. 'Room for the River' has accelerated and enriched the development of nature in Millingerwaard. Diversity has increased with marshes, pastures full of flowers, seepage gullies and alluvial ridges. There is an enormous wealth of different species of river valley plants, fish, birds, insects, mammals and so on. When water levels rise, it is no problem if the whole area is flooded as the riverbank dunes form a safe haven for the animals.

The deserted brick factory is preserved as a relic

Sample Sheet of Intervention – overview on page 24-27

Interplay of city and river

Programme: **Room for the River**
Province: **Gelderland**
Coordinates: **51°51'22.3"N 5°51'25.4"E**
GPS: **51.856194, 5.857056**
Assignment: **increase the flow of the River Waal at Nijmegen**
Implementation: **project decision in 2011 – realisation in 2016**
Water board: **Rivierenland Water Board**
Design engineer(s): **H+N+S Landscape Architects (coordinating architect, landscape design), NEXT Architects (Citadel Bridge), Ney-Poulissen (Promenade Bridge), Trafique, Zwarts Jansma Architects (Extended Waal Bridge)**
Contractor/consortium: **i-Lent consortium consisting of Dura Vermeer Divisie Infra BV and Ploegam BV**
Stakeholders: **province of Gelderland, Rivierenland water board, Directorate-General for Public Works and Water Management, eastern Netherlands division, Arnhem-Nijmegen urban region**
Drop in water level: **34 cm**
Project Area: **250 ha**
Earthworks: **5 million m³**

The Spiegelwaal channel near Nijmegen is far from unique when compared with other projects along the rivers. A piece of cake, really. The River Waal makes a bend here and has too little room. The obvious solution is to reroute the dike and excavate a secondary channel. The same was done at Westenholte in Zwolle and in Munnikenland (see p. 248, 136). That the Spiegelwaal has come to epitomise the new way of dealing with water has everything to do with the urban development that took place at the same time. While hydraulic engineers were figuring out how to give the Waal more room, the urban planners in Nijmegen were making preparations for a 'leap' across the River Waal. By developing the land on the right bank of the Waal, the city would expand towards Arnhem. That would give shape to the concept of the Arnhem-Nijmegen urban region and put the old city centre in a more central position. The ambitions for the river and for the city could be combined, which gave both projects a broader significance. Nijmegen would now be making a triple jump over the Waal, with a new island linking the old city to the new.

Spiegelwaal

Nijmegen is one of the oldest cities in the Netherlands, possibly the oldest of them all. Its location is unique for the Netherlands, with an upper city and a lower city. The

Construction activities for the dike repositioning near Lent in full swing, 2013

The intervention has been completed and the water finds its new way, 2016

Construction day. An elevated work platform provides an overview of the impressive construction site at Lent, 2014

Waal makes a large bend as it rounds the hill on which Nijmegen sits. The room for widening the river was found on the opposite side, next to the village of Lent. Here, the dike was pushed back by 350 m to give room for a second river: the secondary channel, which is fed by culverts carrying water from the Waal when water levels are low and which flows with the Waal when levels rise. The name for this new river, Spiegelwaal ('mirror Waal'), was chosen in a competition among local residents. Jan de Geer from Lent, the lucky winner, won a balloon trip for two people as a result. The Spiegelwaal is 3 km long and up to 5 m deep. A combination of pontoons and poles prevents boats from sailing into the subsidiary channel. The island between the Waal and Spiegelwaal has been christened Veur-Lent. Thanks to this island, the Waal has maintained its historic dimensions in the vicinity of the old part of the city. The embankment lined with buildings at Lent could also largely be spared. The biggest advantage of the island is that it prevents the two riverbanks from being driven apart. Veur-Lent has added an important feature to the centre of Nijmegen in the form of a large river park. This park is easily accessible with three bridges to the old city and three bridges to the new city.

From resistance to participation

The rerouting of the dike and the construction of the Spiegelwaal channel did not go entirely smoothly. The local residents did not immediately see why building an island would make them safer. The village of Lent resisted the changes – the construction of large residential districts and the rerouting of the dikes. Not only would Lent lose its character as a village, but residents would also have to relocate and houses would have to be demolished. A great deal of energy was invested in consultations with all parties to arrive at a solution that, in addition to the unavoidable alterations, would bring as many benefits as possible in the form of amenities, out-of-the-ordinary spots and spatial quality. Residents and businesses were able to make suggestions

2,600 flags mark the trajectory of the future Spiegelwaal, 2012

for the island. That resulted in a cluster of watersports facilities in the Spiegelwaal for rowing, sailing, canoeing and swimming. Other ideas concerned events such as the festivals 'Op 't Eiland' and 'De Oversteek' and the Lent Riverbank Games. There are various experiments in sustainable construction and climate-proof urban extensions.

Granite and bentonite

The new flood defence along the Spiegelwaal consists of embankments cladded with Portuguese granite (the Lentse Warande) and green dikes. Seepage cut-off walls were placed to prevent water seeping through the flood defence ('piping'). The seepage screens consisted partly of sheet pile walls and partly of cement-bentonite that was poured in liquid form into grooves up to 20 m deep. Parts of the island were raised to allow for the construction of houses. The areas that have been laid out for recreation, water and nature are lower lying and are flooded when water levels rise.

Central park as a connecting element

The Spiegelwaal has brought so much more to Nijmegen than just water safety. The city has become more attractive and more diverse. The central park has given the city a green, water-rich heart and brought the new urban expansion within reach. The Waal now flows straight through Nijmegen rather than past it. The activities on Veur-Lent are ideal 'place makers', putting the area and the city expansion on the map. The island is now both in the centre and on the periphery at the same time. The temporary functions and pioneering' adventure are slowly but surely making way for permanent usage. In this way, the new central area is evolving and maturing.

The Lentse Waard with the inlet (Waalval) for the Spiegelwaal. The Waal bridge is in the background, 2016

Nijmegen as a city of bridges

Nijmegen's railway bridge dates from 1880. Three pillars under the access ramp to the River Waal has been reinforced and now stand in the Spiegelwaal. The Waal bridge from 1936 is a classic. Heavy fighting took place in the Second World War in the attempt to control this bridge (one of the ones that featured in *A Bridge Too Far*); as a result, a variety of bombs and grenades had to be removed when the river was widened. The bridge was extended over the Spiegelwaal. The impressive element is under the bridge deck, where the concrete pillars refer to the steel arch of the old main bridge. The design allows for the insertion of a heat exchanger in the floor of the secondary channel, which will provide energy for three hundred planned homes. The leap across the Waal made it necessary to build a third bridge, called De Oversteek ('the crossing'). The same architects, Ney and Poulissen, also designed the new main bridge (the Lentloper or Promenade Bridge) across the Spiegelwaal, which is oriented towards the Sint Stevenskerk, a historic church in the old upper city. The bridge deck consists of a high central section for cars, flanked by lower sections for slow-moving traffic. It introduces a vaulted underside that is supported by well-formed columns. Pedestrian passageways have been created through the vaults to let people experience the Spiegelwaal. The third bridge over the Spiegelwaal is called the Zaligebrug (Blessed Bridge) or Citadel Bridge; it takes cyclists on a meandering route across the water.

Beautifully shaped columns carry the Lentloper; in the background the railway bridge and the Oversteek

The Waalbrug that was extended over the secondary channel

Underpass of the Lentloper

Snelbinder stairs, a connection for pedestrians and cyclists between the Snelbinder and the Veur-Lent island

Pedestrian passage under the Lentloper

03 Sample Sheet of Intervention – overview on page 24-27
14
15

Switching between the town and the countryside and between culture and nature

Project: **Keizerswaarden, Stobbenwaarden and Olsterwaarden (east riverbank) and Bolwerksplas, Worp and Ossenwaard (west riverbank)**
Programme: **Room for the River**
Province: **Overijssel**
Coordinates: **52°16'46.9"N 6°06'59.8"E**
GPS: **52.279695, 6.116609**
Assignment: **increase the flow of the River Ijssel at Deventer**
Implementation: **project decision in 2010 – work started in 2012 – flood protection delivery in 2015**
Water board: **Drents Overijsselse Delta water board (east riverbank), Vallei en Veluwe water board (west riverbank)**
Design engineer(s): **Layout plan: Royal Haskoning/VHP, Natuurderij: Paridon & de Groot (country estate) and Daad Architects (farm)**
Contractor/consortium: **IJsselfront consortium (Boskalis/Van Hattum en Blankevoort), consultancy by Witteveen + Bos.**
Stakeholders: **Municipality of Deventer, Provincial authority of Overijssel, IJssellandschap Foundation, watersports associations, ProRail**
Drop in water level: **10 cm (Keizerswaarden, Stobbenwaarden and Olsterwaarden) and 18 cm (Bolwerksplas, Worp and Ossenwaard)**
Area: **500 ha**
Earthworks: **2.7 million m³ for the excavation of the floodplains**

Deventer has an imposing waterfront along the River IJssel. Traces of the mediaeval town wall can be seen in it, and of course the Lebuïnuskerk, the historic church that dominates the wider area. The opposite side of the IJssel is largely a river landscape although it also contains an urban park (the Worpplantsoen) and a residential district.
In the 1960s there were plans to make Deventer a two-centre city straddling the river. But there was a lot of opposition to this. Fortunately, Deventer preserved the striking contrast between the urban riverbank and the rural riverbank. The IJssel does not have much room at the point where it passes the old town and the Worpplantsoen. Deventer has a long history of high water levels, water washing over embankments and flooded cellars. The floodplains on either side of the river have been excavated in an effort to boost water flows through this bottleneck and reduce the risk of flooding.

Differentiation in the floodplains

New river arms were created in the lower lying parts of the floodplains on both banks. These are channels that come to a dead end in the floodplains. Three such arms were dug on either side: they form virtually continuous secondary channels. The floodplains here are rough, dynamic and varied – ideal for the nature associated with rivers. A few facilities were introduced in some places, such as paths, spots for angling and a birdwatching hut. The higher lying parts of the floodplains have been brought into cultivation more, either as farmland or for recreational purposes. This has resulted in differentiation between the two riverbanks and between the individual forelands. The old Worpplantsoen opposite the town centre was given a facelift. The embankment, built from natural stone, has been raised slightly and a green slope has been added that fits well with the park and raises the height further still. At the point that used to be the start of a bridge resting on boats, a ferry now picks up cyclists and pedestrians to take them across to the city. Watersports activities are clustered around the Zandweerdplas lake, with a marina, a rowing lane, a floating boathouse and the watersports association clubhouse.

High water on the IJsselkade in Deventer, 2011

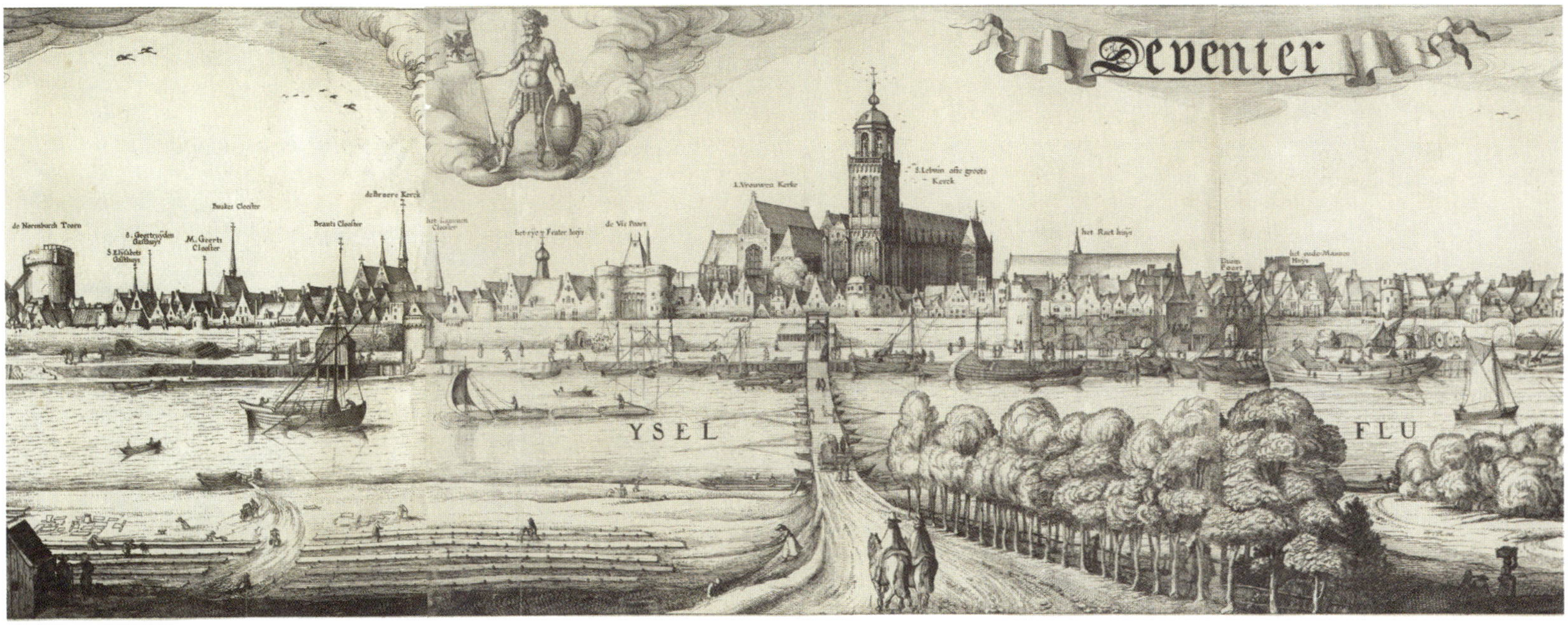

'Profiel van Deventer' (1615), etching by Claes Jansz. Visser

The view of Deventer today from the jetty at De Worp

Natuurderij

A new country estate, the Natuurderij, was created outside the dike in the Keizerswaard and Stobbenwaard, just to the north of the city. IJssellandschap Foundation built the complex and is leasing it. It is actually a minor miracle that the enterprise was built at all. New building outside the dikes and next to major rivers is discouraged and the establishment of a farm in a nature area is usually quite out of the question because of the ammonia emissions. The Natuurderij is the first Dutch undertaking to combine organic farming, nature management and water management. Its aim is to be self-sufficient, increase biodiversity and conserve typical landscape elements. It is sticking closely to the management guidelines for nature areas (Natura 2000). The cows graze in the floodplain, which keeps the landscape open. The farm and its barns, which were designed by Daad Architects, are located on a mound with superb views. The mound is surrounded by a concrete defence wall. The Natuurderij has become a popular destination for Deventer's inhabitants thanks to its location just outside the city and close to historic residences that are also partially open to the public.

Urban overflow area

Deventer is (once again) closely connected to the river thanks to the Worp, the watersports facilities at Zandweerdplas, the agricultural estate of Natuurderij and the many cycling and hiking routes. The city's inhabitants can stroll along a boulevard, roam through the countryside, walk along the cliff-edge path, cycle around the Keizerslanden or do a circuit over the IJssel and back via the Wilhelmina Bridge, Worpplantsoen and the railway bridge. In addition to urban overflow areas, the intervention has produced nature areas (Hengforderwaarden and Ossenwaard) and safeguarded the use of the land for agriculture. An almost continuous 100 m wide secondary river has been created alongside the city. The floodplains on the left bank form a continuous ribbon of green, with natural grasslands and the occasional riparian woodland.

Floating jetties in the Zandweerdplas near Deventer

High water in the IJssel near Deventer now has more room so that the quay remains dry, 2016

Walking along the secondary channel towards Natuurderij Keizersrande

The Keizerswaarden and Hengforderwaarden before the intervention, 2012

The Keizerswaarden and Hengforderwaarden during the intervention, 2013

The Keizerswaarden
and Hengforderwaarden
after the intervention

Sample Sheet of Intervention – overview on page 24-27

Like a painting by Jan Voerman

Programme: **Room for the River**
Province: **Overijssel**
Coordinates: **52°30'52.1"N 6°01'55.9"E**
GPS: **52.514460, 6.032201**
Assignment: **give the River IJssel more room when water levels rise and improve the spatial quality**
Implementation: **2011-2016**
Water board: **Drents Overijsselse Delta water board (Groot Salland water board before 2016)**
Design engineer(s): **Bosch Slabbers in collaboration with Arcadis, IPV Delft (floodable bridge)**
Contractor/consortium: **Van den Biggelaar / Ploegam**
Stakeholders: **municipality of Zwolle, Natuurmonumenten, Staatsbosbeheer, Directorate-General for Public Works and Water Management, private individuals (residents living on floodplains), country estates, ProRail**
Drop in water level: **14 cm (Westenholte) and 9 cm (Scheller and Oldeneler Buitenwaarden)**
Project Area: **not known**
Earthworks: **1.3 million m^3**

Like Deventer and Kampen, Zwolle is a centuries-old Hanseatic town on the River IJssel. It lies only about 20 km from the mouth of the river in Ketelmeer, a tongue of the IJsselmeer Lake. Like Kampen, Zwolle is also at risk whenever high water levels in the River IJssel coincide with a north-westerly storm pushing up the water in the IJsselmeer Lake (X1). When this happens, water comes at the town from two sides and has nowhere to go. The risk of flooding has been substantially reduced by giving the river at Zwolle more room in a number of different ways. The dike to the north of the city near the Westenholte district was moved back over a length of more than 2 km. That made the river bed about 300 m wider.

Widening, deepening and doubling

A second alteration was to make the Scheller and Oldeneler Buitenwaarden deeper on the south side of the city. As much as 4 m was dug out in some places. The third measure involved deepening the IJssel's summer bed between Zwolle and Kampen to let the river discharge more water. In the process, a bypass was created behind Kampen to give the IJssel a reserve mouth in Drontermeer. A unique find was made near Kampen during the preparations for the dredging work in the IJssel: a well-preserved cargo ship from the Middle Ages. This vessel, of the type known as cogs, lay hidden 4 m below the riverbed surface with only one side protruding slightly above the surface. It was raised up as part of the Room for

Bridge over the flood opening of the Strang

The picturesque River IJssel

Jan Voerman (1857-1941) was a farmer's son from Kampen. After training at the Dutch Royal Academy for the Fine Arts, he set up shop as an artist in Amsterdam. He painted a great many landscapes and watercolours of flowers. After his marriage to Anna Verkade (in 1889) he moved to Hattem, near Zwolle. From that time on, the landscape of the IJssel became an important theme in his work. His son Jan (1890-1976) was also an artist. He became famous in the Netherlands for his illustrations for the albums produced by the Verkade food company, which started life as a biscuit factory run by his grandfather. One album after another was published between 1906 and 1939. They made generations of Dutch people aware of the beauty of their country and its nature. The work of both father and son can be seen in the Voerman Museum in Hattem.

The river was a favourite subject for the painter Jan Voerman sr. The designers of today draw inspiration from his views of the river

The floodplain area at Westenholte before the intervention, 2010

The same floodplain area after excavation works

During high water this 200-metere-long bridge disappears under water

the River project and is now carefully being restored in the Batavia wharf in Lelystad. Such finds offer an incomparable insight into life in days gone by – in this case, a journey through time to a treasure chamber from the fifteenth century.

Modern programme in an idyllic setting

The design for the rerouting of the dike at Westenholte also began as a journey back in time. The assignment involved the construction of a new dike, digging a secondary channel in the floodplains and laying out the area outside the dike. The landscape architects Bosch Slabbers took their inspiration from the historical appearance of the river areas. In the notes accompanying their design, they write that they were inspired by the views of the IJssel painted by the artist Jan Voerman in around 1900. His paintings show idyllic scenes with cows in the floodplains, winding riverbanks and the outline of a village and copses on the horizon. This image is reflected in the layout of the secondary channels that have been excavated in the floodplains. However, the area outside the dike also features numerous elements that Voerman would never have encountered in his trips along the IJssel. They include a bird-watching hut, a 200 m long bridge that disappears under water when water levels rise and the new country estate of Vreugderijk on top of the dike with terraces on the river side that become submerged one by one as the river rises.

Terraces and a floodable bridge

The new dike is high but the visual effect is tempered by the projecting terraces that have been placed on the river side. There is room here for strips of woodland and one or two homes. Three dike houses have been designated for local inhabitants who were forced to move home because of the rerouting of the dike. The new country residence of Vreugderijk is currently being developed. The island created by the construction of the secondary channel is connected to the outside world by two bridges. Three design studios were invited to submit a proposal. The project office asked for an attractive but sober design that would fit well with the existing landscape. Sixty votes were cast in a survey of residents in the nearby districts of Spoolde and Westenholte. Of the two preferred alternatives, the supervisory committee chose a design by IPV Delft. The bridge is low and austere, its dimensions designed to withstand the force of the river water. The materials have been chosen to cope with flooding. The area is close to Vreugdehoeve, which is managed by Natuurmonumenten, the Dutch society for the preservation of nature.
It offers the ideal combination of nature development and recreational development in the area outside the dike.

Godwits, great egrets and ruddy shelducks can be viewed from this bird observation hut

Residents who had to make way for the interventions built new houses along the dike

These residents also retained their splendid views

Acknowledgments

Interviews

Jos van Alphen and Liliane van Sprundel, 18 May 2017
Ingwer de Boer, 26 July 2017
Richard Jorissen, 9 May 2017
Michel Lascaris, Ellen Vreenegoor, Barbara Speleers and Gerard Troost (RCE), 4 July 2017
Inge Lardinois, 12 July 2017
Henk Ovink, 29 May 2017
Hein Pieper, 21 June 2017
Jan Terlouw, 8 June 2017
Willie Tiggeloven, 26 July 2017
Dirk Sijmons, 3 May 2017
Sylvo Thijsen, 22 June 2017
With special thanks to Tjalle de Haan for his invaluable comments on the essay.

Bibliography

Angremond, K. d', P. Huisman, T. de Jong, G.J. Schiereck, W. Thissen et al., *Watertovenaars. Delftse ideeën voor nog 200 jaar* (Rotterdam: Rijkswaterstaat, 1998)

Beek, Marijke, and Marinus Kooiman, *Aan de dijk gezet. Dijkwoningen vroeger, nu en in de toekomst*, (Rotterdam: 1996)

Biesboer, F., 'Ties Rijcken over de worsteling van de waterbouw: ons watersysteem verdient schoonheid', *De Ingenieur*, June (2017)

Bijhouwer, J.T.P., 'Behouden of scheppen', *Tijdschrift voor Volkshuisvesting en Stedenbouw* 15 (1934) 3, pp. 34–35

Cantrell, B., and Justine Holzman, *Responsive Landscapes* (Abingdon, 2016)

Bruin, D. de, D. Hamhuis, L. van Nieuwenhuijze, W. Overmars, D. Sijmons, F. Vera *Plan Ooievaar. Een toekomst voor het rivierengebied*: Gelderse Milieufederatie, Rijkswaterstaat, Ministerie van LNV, (1987)

Bruin, D. de, *Wat Er Staat: Civiele techniek en ruimtelijke ordening* (Ministerie van VROM, 2007), pp. 85–107

Buuren, Arwin van, G. R. Teisman, J. Verkerk and M. Eldering, *Samen verder werken aan de Delta. De governance van het Nationaal Deltaprogramma na 2014* (Rotterdam: Erasmus Universiteit, 2014)

Buuren, A. van, G. J. Ellen, C. van Leeuwen, J. van Popering–Verkerk, *Die het water deert die het water keert. Overstromingsrisicobeheer als maatschappelijke gebiedsopgave. Opbrengsten en lessen uit de pilots meerlaagsveiligheid* (Rotterdam: Erasmus Universiteit, 2015)

Dieren, W. van, *Organogene Dünenbildung, eine geomorphologische Analyse der westfriesischen Insel Terschelling mit pflanzensoziologischen Methoden* (1934)

Dijk, T. van, 'De lus is verbroken' (interview with Professor Han Vrijling on his retirement], Delft Integraal (2013) 3, pp. 10-12

Dooren, N. van, and G. van der Ven, *De nieuwe rivierdijken. Dijkversterking als ontwerpopgave*, (Rotterdam: Nederlands Architectuurinstituut, 1997)

Eijndhoven, J. van, M. El Filali, A. van Hall, H. Hooghoff, H. Pleij, D. Sijmons, B. Toussaint and R. van der Vaart, 'Zoden aan de dijk. 25 peilingen naar Nederland als waterland', Amsterdam (2009)

Feddes, Y.C., F.L. Halenbeek, *Een scherpe grens; ontwerpstudie naar de ruimtelijke kwaliteit van verzwaarde rivierdijken* (Utrecht: Staatsbosbeheer 1988)

Gast, C. de, *De macht van het water. Leven met water tussen Maas en Merwede*, (2004)

Geworteld in het verleden, op koers naar de toekomst. Samenwerkingsovereenkomst Rijkswaterstaat-Staatsbosbeheer 2017-2020

De grote rivieren in beeld. De oogst van twintig jaar natuurontwikkeling en uiterwaardinrichting langs de Rijntakken en de Maas (2012)

Heezik, A. van, *Strijd om de rivieren. 200 jaar rivierenbeleid in Nederland of de opkomst en ondergang van het streven naar de normale rivier* (Haarlem/The Hague, 2008)

Hellman, Paul, 'De nadagen van het rivierenland. Het vergeefse protest tegen dijkverzwaring', *NRC Handelsblad*, Saturday supplement (3 September 1994), p. 3

Hindriksma, Martin, 'Dijkbewoners baas over eigen dijk', *Binnenlands Bestuur*, 10 July 2017

'Kamer en regering oneens met milieu tot zondebok te maken. Jorritsma: verwaarlozen dijken kwam door de politiek', *Amigoe*, 4 February 1995

Kaufman-Osborn, T., *Creatures of Prometheus: Gender and the Politics of Technology* (Lanham 1997)

'Kennissysteem LNV hobbelt achter innovaties aan', *Wetenschap*, 14 November 1996

Kimmelman, M., 'The Dutch have solutions to rising seas. The world is watching', *The New York Times* (15 June 2017)

Lokman, K., 'Cyborg landscapes: Choreographing resilient interactions between infrastructure, ecology and society', *Journal of Landscape Architecture* 12 (2017) 1, 60–73

Lörzing, Han, 'Ooievaar. De toekomst van het rivierengebied', *NRC Handelsblad* (16 July 1987)

Mazzucato, Maria, *The Entrepreneural State: Debunking Public vs. Private Sector Myths* (London: 2013)

'Minister May positief over dijkrapport', *Nieuwsblad van het Noorden* (12 January 1993)

Mensing, M., 'Reputatie op het spel bij bedwingen Ganges', *De Telegraaf* (18 December 1985)

Metz, T., and M. van den Heuvel, *Zoet en zout. Water en de Nederlanders* (Rotterdam: 2012)

Milikowski, F., 'Klotsende golven tegen de superdijk', *De Groene Amsterdammer* 20, May 2015, pp. 22-27

'Millingerwaard' op website ARK, (accessed 15 July 2017)

OECD report with link: http://www.oecd.org/cfe/regional-policy/water-governance-netherlands.htm

'Onderzoek laakt dijkverzwaring', *NRC Handelsblad*, 11 januari 1993

Projecten 2017. Hoogwaterbescherming Projectenboek, Programmadirectie Hoogwaterbescherming, LOLA landscape architects, Breda, November 2016

Ridder, H.A.J. de, 'Van maker naar regisseur. Een nieuwe rol voor Rijkswaterstaat', in: K. d'Angremond, P. Huisman, T. de Jong, G.J. Schiereck, W. Thissen et al, *Watertovenaars. Delftse ideeën voor nog 200 jaar Rijkswaterstaat* (Rotterdam: 1998), pp. 305–309

Roncken, P., S. Stremke, R. M. Pulselli, 'Landscape Machines Designerly Concept and Framework for an Evolving Discourse on Living System Design', in: D. Czechowski, T. Hauck, G. Hausladen, *Revising Green Infrastructure: Concepts Between Nature and Design* (2014) pp. 91–112

Ruiter, F.G. de, 'Rumoer in rivierenland. Verhoging van de dijken bedreigt twaalf karakteristieke huisjes in Neerijnen aan de Waal. In Sliedrecht aan de Merwede kan het wel; behoud van dijkhuizen', *NRC Handelsblad* 10 January 1987

Rijcken, Ties, *EMERGO, the Dutch Flood Risk System Since 1986*, doctoral thesis Delft University of Technology (2017) PhD supervisor Prof.drs.ir. Han Vrijling (CiTG) and Prof.dr.ir. Matthijs Kok

Saeijs, H.L.F., 'Conditioneren. Lessen uit 20 eeuwen waterbouwkunde en omgaan met water', *Driemaandelijkse Deltaberichten* 123/124 (1988), pp. 695-716

Santbergen, H., *Henk Saeijs. Stormloper in een delta*, Delft (2015)

Sijmons, D.F., 'Bewogen Beweging: Landschap onder de kap', Farewell speech Delft University of Technology, 16 October 2015

Steenhuis, M., and F. Hooimeijer (ed.), *Maakbaar Landschap. Nederlandse landschapsarchitectuur 1945-1970* (Rotterdam: 2009)

Steenhuis, M., 'Overbodig zondebesef : de emancipatie van de naoorlogse landschapsarchitectuur', Bijhouwer Lecture, Wageningen (2015)

Steenhuis, M. (ed.), *De Deltawerken* (Rotterdam: 2016)

Stumpe, J., F. Tielrooy, *Geef water de ruimte en aandacht die het verdient, Advies van de Watercommissie 21e eeuw* (2000)

Tonk, F., Richard Jorissen (Hoogwaterbeschermingsprogramma): Innovatie is keihard nodig', website Neerlands Diep, Academie voor Publiek Bouw- en Infraprojecten, (accessed 24 July 2017)

"'Uiterwaarden moeten zonodig onteigend' Smit-Kroes wil plan Ooievaar doorzetten" *Digibron*, 13 October 1989

Velden, Ben van der, 'Hoogwater in Waal door Deltawerken. Kostbare extra dijkverhogingen moeten negatieve gevolgen beperken', *NRC Handelsblad* (2 October 1984), 1.

Velden, Ben van der, 'Waterstaat soms feilbare rekenmeester. Niet alle gevolgen van Deltaplan zijn in jaren 50 voorzien', *NRC Handelsblad* (2 October 1984), 3

Vrijling, J.K., and J. Stoop, 'Naar één beslismodel voor de veiligheid. Rijkswaterstaat en de kansberekening', in: K. d'Angremond, P. Huisman, T. de Jong, G.J. Schiereck, W. Thissen et al. *Watertovenaars. Delftse ideeën voor nog 200 jaar Rijkswaterstaat* (Rotterdam:, 1998), pp. 202–213

Vroom, M.J., *Leren kijken. Het Wageningse onderwijs en onderzoek in de tuin- en landschapsarchitectuur* (Wageningen: 2014), pp. 175–176

website: https://www.ark.eu/over-ark/ark-organisatie/25-jaar-ark/25-jaar-ark-gelderse-poort

Wijkman, Anders, and Johan Rockström, Strategies for planetary stewardship are desperately needed, *The Guardian*, 6 December 2012

Woud, A. van der, 'De zeedijken', in: *Het Lege Land. De ruimtelijke orde van Nederland 1798-1848* (Amsterdam: 1987), pp. 80–91

Image credits

Photography

a above, b below, r right, l left
ADC ArcheoProjecten 171
Als de dijken breken 46
Arcadis 211b
Bert Beelen / Hollandse Hoogte 234a, 236
Gerlo Beernink / Hollandse Hoogte 53
Henk Braam / Hollandse Hoogte 108a
Els Branderhorst / Buiten-Beeld 230
Peter Buteijn 85b
Collectie familie Lardinois 49al, 49ar
Consortium Grensmaas 162a (Archeodienst), 166/167 (Flying Eye)
Henri Cormont 12/13, 110/111, 116a, 116b, 117a,117b, 132, 147a, 148a, 182al
Delfstoffen Combinatie Maasdal (DCM) 168
Deltares 51
Tineke Dijkstra cover, first endpaper 2/3, 16, 32/33,39a, 41, 50, 57b, 60/61, 64/65, 67r, 68, 69, 71b, 72, 73a, 73b, 74, 76/77, 79a, 81, 88, 89, 91b, 92, 93, 94r, 95l, 95r, 98b, 99, 101al, 101bl, 102a, 102b, 104/105, 112/113, 114/115, 122, 123ar, 124/125, 130, 131b, 135a, 136b, 139, 140, 144a, 144b, 145ar, 147b, 148b, 152/153, 154/155, 159, 161b, 163, 164, 165, 170, 173a, 173b, 174a, 176/177, 182ar, 182 br, 183, 185, 186, 187, 188, 189, 190b, 196, 197al, 197bl, 197ar, 197br, 198a, 198b, 199, 201a, 201b, 203, 218/219, 222, 223, 224, 225, 226, 227, 228a, 239, 240al, 240bl, 240r, 241a, 241b, 243b, 244, 245a, 248, 251, 252, 253a, 253b, second endpaper
Roger Dohmen / Hollandse Hoogte 134, 158r
Gelderse Natuur en Milieufederatie 229a
Gemeente Bergen, Limburg 228b
Flip Franssen / Hollandse Hoogte 75, 184a, 184b, 214/215, 234b,
John Gundlach first endpaper / 1
Sijmen Hendriks / Hollandse Hoogte 229b
Peter Hilz / Hollandse Hoogte 37, 242
Hoogheemraadschap Hollands Noorderkwartier 36
Caspar Huurdeman / Hollandse Hoogte 208ar
Vincent Jannink / ANP 161al
Bert Janssen / Hollandse Hoogte 156
Marcel Kentin 94l
Chris Keulen / Hollandse Hoogte 160, 162b
Jaco Klamer / Hollandse Hoogte 216/217
Marcel Köppen / Beeldkazerne 136b
Olaf Kraak / Hollandse Hoogte 82
Luuk Kramer 101br
John Lambrichts / Hollandse Hoogte 158l
Frans Lemmens / Hollandse Hoogte 84, 107b
Ger Loeffen / Hollandse Hoogte 49b
Marco van Middelkoop / Hollandse Hoogte 86/87, 178/179, 221a, 237b
Ministerie van Infrastructuur en Milieu / Jentse Hoekstra 192, 193
Ministerie van Infrastructuur en Milieu / Willem Mak 212a
Phil Nijhuis / Hollandse Hoogte 98a
Wouter Pattyn / Hollandse Hoogte 150/151
Chris Pennarts / Hollandse Hoogte 127
Bram Petraeus / Hollandse Hoogte 208al, 208br
Rebuild by design 212b
Rijkswaterstaat / Werry Crone 58a, 107b, 126, 145b, 74b, 235, 245b
Rijkswaterstaat / Joop van Houdt 66, 85a, 91a
Rijkswaterstaat / Marcel Köppen 137a
Rijkswaterstaat / Martin van Lokven 231a
Rijkswaterstaat / Marcel Molle 10/11
Freddy Schinkel 180/181, 204, 205a
Siebe Swart 6/7, 35, 39b, 42, 55, 57a, 62/63, 67l, 71a, 80, 83, 96, 100, 103, 108b, 118/119, 120, 123al, 123b, 128/129, 131a, 133, 135b, 137b, 138, 141, 142/143, 157, 161ar, 169, 190a, 194/195, 206, 209a, 220, 221b, 232/233, 237a, 247, 250a, 250b
SteenhuisMeurs 40
George Steinmetz / Hollandse Hoogte 211a
Twan Teunissen / ARK Natuurontwikkeling 231b
Richard Trenning 78, 79b
Koen Verheijden / Hollandse Hoogte 209b, 238
Waterschap Drents Overijsselse Delta 202 (John Maaskant), 205b (Hoogwaterbrigade) 246a, 246b (Bram van de Biezen)
Waterschap Noorderzijlvest 90
Waterschap Rivierenland / August Swietkoviak 45o
Waterschap Vallei en Veluwe 208bl
Ries van Wendel de Joode / Hollandse Hoogte 145al
Wetterskip Fryslân / Daniel Hartog 70

Other Images

15 Jos Dirksen
23b Rikkers Infographics
35b map of Amsterdam, source: www.atlasnatuurlijk-kapitaal.nl
45a Plan Ooievaar concept sketch
58b OKRA Landschapsarchitecten artist impression Katwijk
97 Hendrik Willem Mesdag Panorama Mesdag (detail) 1881 Panorama Mesdag, The Hague
121 Doorvaart bij Pannerden, 1701, source: Rijkswaterstaat
191 Indijking Friesche Wadden, source: Collectie Nationaal Archief
200 L. Rohbock and A. Fesca Kampen, engraving ± 1870, source: Collectie Zuiderzeemuseum (image bank Cultureel Erfgoed)
207 Johann van den Noort sketch
212 Rebuild by Design
243b Claes Jansz. Visscher Profiel van Deventer (plate II) 1615, source: Rijksmuseum
249b Jan Voerman, IJsselgezicht, zomeravond bij opkomend onweer, oil paint on panel, source: Voerman Museum Hattem
249b Jan Voerman, Gezicht op de IJssel, heldere dag, oil paint on panel, source: Voerman Museum Hattem (in loan of Dordrechts Museum)

Credits

Text essay and objets trouvés: **Marinke Steenhuis**
Text sample sheet of interventions and themed texts: **Paul Meurs**
Text projects: **Paul Meurs and Annet Kampinga**
Project guidance: **Kees Poot, Jentse Hoekstra**
Translation: **Vivien Reid and Clare Wilkinson**
Text editing: **Bookmakers**
Image editing: **Tineke Dijkstra, Paul van Hamersveld**
Design: **Beukers Scholma**
Photography: **Tineke Dijkstra, Siebe Swart**
Cartography and infographics: **Mijs Cartografie en Vormgeving**
Lithography and printing: **NPN Drukkers**
Publisher: **Marcel Witvoet, nai010 publishers**

This publication was made possible by the Ministry of Infrastructure and Environment.

This book was printed on FSC certified paper. For the cover Gmund f-color Metallic 310 micro crystal 130 g was used. For the endpapers and the inside Edixion offset 150 g was used.

nai010 publishers is an internationally orientated publisher specialized in developing, producing and distributing books in the fields of architecture, urbanism, art and design. www.nai010.com

nai010 books are available internationally at selected bookstores and from the following distribution partners:
North, Central and South America
Artbook | D.A.P., New York, USA
dap@dapinc.com
Rest of the world
Idea Books, Amsterdam, the Netherlands
idea@ideabooks.nl

For general questions, please contact nai010 publishers directly at sales@nai010.com or visit our website www.nai010.com for further information.

Printed and bound in the Netherlands

ISBN 978-94-6208-384-4

Beyond the Dikes is also available as an e-book and in Dutch:

Beyond the Dikes e-book **(pdf)**
ISBN 978-94-6208-403-2
Voorbij de dijken
ISBN 978-94-6208-383-7
Voorbij de dijken e-boek **(pdf)**
ISBN 978-94-6208-402-5

Brakelse Benedenwaarden and Buitenpolder Munnikenland – Spotting birds (including spoonbills) in the Brakelse Benedenwaarden along the Waal